Springer Tracts in Mechanical Engineering

About this Series

Springer Tracts in Mechanical Engineering (STME) publishes the latest developments in Mechanical Engineering - quickly, informally and with high quality. The intent is to cover all the main branches of mechanical engineering, both theoretical and applied, including:

- Engineering Design
- Machinery and Machine Elements
- Mechanical structures and stress analysis
- Automotive Engineering
- Engine Technology
- Aerospace Technology and Astronautics
- Nanotechnology and Microengineering
- Control, Robotics, Mechatronics
- MEMS
- Theoretical and Applied Mechanics
- Dynamical Systems, Control
- Fluids mechanics
- Engineering Thermodynamics, Heat and Mass Transfer
- Manufacturing
- Precision engineering, Instrumentation, Measurement
- Materials Engineering
- Tribology and surface technology

Within the scopes of the series are monographs, professional books or graduate textbooks, edited volumes as well as outstanding PhD theses and books purposely devoted to support education in mechanical engineering at graduate and post-graduate levels.

More information about this series at
http://www.springer.com/series/11693

Bernhard Eisfeld
Editor

Differential Reynolds Stress Modeling for Separating Flows in Industrial Aerodynamics

Editor
Bernhard Eisfeld
German Aerospace Center (DLR)
Institute of Aerodynamics and Flow
Technology
Braunschweig
Germany

ISSN 2195-9862 ISSN 2195-9870 (electronic)
Springer Tracts in Mechanical Engineering
ISBN 978-3-319-38569-3 ISBN 978-3-319-15639-2 (eBook)
DOI 10.1007/978-3-319-15639-2

Springer Cham Heidelberg New York Dordrecht London

Softcover reprint of the hardcover 1st edition 2015

Printed on acid-free paper

Springer International Publishing AG Switzerland is part of Springer Science+Business Media (www.springer.com)

Preface

Flow separation has attracted fluid mechanics research for a long time, whereas in industrial aerodynamic design, flow separation is usually avoided due to its detrimental effect on the performance of the respective apparatus. Flow separation often limits a machine's performance, e.g. in terms of maximum aircraft lift or in terms of the surge limit of turbo compressors. Also aerodynamic shape optimisation is usually bound by the onset of separation.

Currently there is a trend in industry to rely more and more on data obtained from numerical flow simulations, using Computational Fluid Dynamics (CFD) software. Such CFD-based aerodynamic design therefore heavily depends on the accuracy with which separating flows can be predicted. As long as the flow stays laminar, there is probably less doubt in the reliability of CFD predictions, even of separating flows, but industrially relevant flows are usually turbulent.

In principle, in turbulent flow the governing equations for laminar flow are still valid and could be solved. However, such Direct Numerical Simulations (DNS) need to accurately resolve the turbulent small-scale motion to get the mean flow of interest right, thus demanding computational resources that are currently not affordable in an industrial environment. Resolving only part of the spectrum of turbulent fluctuations by Large Eddy Simulation (LES) relaxes the computational requirements, but still is too expensive for an industrial application of boundary-layer dominated flow at high Reynolds number.

For this reason, the old-fashioned approach based on the Reynolds-averaged Navier–Stokes (RANS) equations and employing corresponding turbulence models will remain the standard technology in most applications of industrial aerodynamics in the next decade(s). Unfortunately, today's standard models, mainly based on the assumption of a turbulence-generated eddy-viscosity, are considered notoriously unreliable for the prediction of separated flows. Therefore there is urgent need for improvement.

One possible path towards an improved prediction of separated flows consists in directly solving the transport equations for the individual Reynolds stresses instead of employing a simplified model based on the assumption of an eddy viscosity. These so-called Differential Reynolds Stress Models (DRSM) constitute the highest level of RANS-based turbulence models. They are certainly not a panacea *per se*, but offer more possibilities for modelling individual effects of turbulence on a higher level according to the physics. In particular the production of turbulence is defined exactly in terms of known quantities.

In general, DRSMs are considered much more difficult to handle in a numerical flow solver, in particular when applied to industrially relevant flows. The contributions in this book, nevertheless, demonstrate their applicability to separated flows with different numerical flow solvers that are developed not only at universities but also at research labs and a company, aiming at industrial use in external aerodynamics as well as in turbomachinery. Moreover, the applications are not restricted to one particular model only, but cover a variety of DRSM flavours, allowing for cross-comparisons.

As might be expected, DRSMs demonstrate advantages when vortices or anisotropy-driven secondary flows are involved. In other cases improvements are not always evident compared to established eddy-viscosity models. Nevertheless, DRSMs perform rarely worse, and since their application still appears to be in a pioneering state, the presented results are considered encouraging for further research in differential Reynolds stress modelling for separating flows in industrial aerodynamics.

Braunschweig, Germany
November 2014

Bernhard Eisfeld

Contents

Application of a Low Reynolds Differential Reynolds Stress Model to a Compressor Cascade Tip-Leakage Flow

Christian Morsbach, Martin Franke, and Francesca di Mare

Abstract The tip-leakage flow of a low speed compressor cascade at $Ma = 0.07$ and $Re = 400{,}000$ was simulated employing the Jakirlić/Hanjalić-ω^h (JH-ω^h) differential Reynolds Stress model (DRSM) and results are presented. The predictions are compared with those obtained using the SSG/LRR-ω DRSM and the Menter SST k-ω linear eddy viscosity model (LEVM). In addition to the mean flow quantities, the focus is on the Reynolds stresses and their anisotropy. Both DRSMs show significant improvements compared to the LEVM with respect to the mean flow quantities; however, details of the turbulence structure are more accurately predicted by the JH-ω^h model.

1 Introduction

The flow in axial compressor rotors is highly complex due to, amongst other phenomena, the vortical motions which develop in the gap between the blades and the machine's casing (tip-gap) [16]. Numerical simulations of such flows often rely on highly tuned linear eddy viscosity models (LEVM) despite the high anisotropy which characterises the turbulence field and plays a major role in many phenomena of practical interest. It would appear sensible to adopt, in these cases, an anisotropy-resolving modelling approach. Differential Reynolds stress models (DRSM) belong to this class of closures; however, reports on their application to realistic configurations are scarce.

Gerolymos and co-workers were among the first to employ DRSMs to investigate complex configurations, ranging from cascades [7] to multi-stage compressors [6].

C. Morsbach (✉) • F. di Mare
Department of Numerical Methods, German Aerospace Center (DLR), Institute of Propulsion Technology, Linder Höhe, 51147 Cologne, Germany
e-mail: christian.morsbach@dlr.de; francesca.dimare@dlr.de

M. Franke
Department of Numerical Methods, German Aerospace Center (DLR), Institute of Propulsion Technology, Müller-Breslau-Str. 8, 10623 Berlin, Germany
e-mail: martin.franke@dlr.de

B. Eisfeld (ed.), *Differential Reynolds Stress Modeling for Separating Flows in Industrial Aerodynamics*, Springer Tracts in Mechanical Engineering,
DOI 10.1007/978-3-319-15639-2_1

They developed a DRSM with special focus on independence from geometry related parameters such as distance from the wall and wall normal vectors. They compared performance data as well as radial distributions of flow angles, total pressure and temperature, etc. to results obtained by a standard k-ϵ approach. While they could show only marginal improvements over LEVMs using DRSMs for flows which are not dominated by large separation, the flows dominated by large separation were predicted in better agreement with experimental data [6]. Rautaheimo [25] conducted simulations of a centrifugal compressor using a DRSM combined of a high and a low Reynolds model. He found that the LEVMs were superior in predicting the integral values whereas the DRSM performed better in regions with secondary flows. A linear compressor cascade with tip-clearance was investigated by Borello et al. [2] using the DRSM of Hanjalić and Jakirlić [9]. In all the above-mentioned studies, results obtained using DRSMs were found to be superior to those of an LEVM taken as reference, especially for complex 3D flow features with anisotropic turbulence. Yet, despite the obvious advantages of DRSMs, they are still not popular in industrial design applications.

In a previous paper, the present authors applied the SSG/LRR-ω DRSM to a compressor cascade flow and compared the results to those obtained with an LEVM and an explicit algebraic Reynolds stress model [22]. It could be shown that the prediction of secondary velocities in the tip-gap flow and the shape of the tip-gap vortex could be improved by the DRSM. However, there was still potential for improvement in the representation of the mean velocities and especially of the Reynolds stresses near the wall. This motivated the present investigation using the low Reynolds DRSM of Jakirlić and Hanjalić in a formulation employing the specific homogeneous dissipation rate ω^h as scale determining variable [18].

2 Turbulence Modelling

In a Reynolds averaged Navier-Stokes (RANS) framework, the objective of turbulence modelling is to determine the Reynolds stress tensor $\overline{\rho u_i'' u_j''}$. This can be accomplished using closures entailing different levels of complexity. For standard industrial CFD applications, the Boussinesq approximation is generally invoked, which defines a turbulent viscosity μ_T to relate the Reynolds stresses directly to the trace-free rate of strain S_{ij}^*. A prominent example of such an approach is the Menter SST k-ω model [20], which will be used as reference in this paper. However, although the linear stress-strain coupling can be justified for certain flow topologies, it cannot be expected to hold in general. In fact, it is the reason for the inability of LEVMs to predict higher order effects such as streamline curvature, rotation or three dimensional boundary layers [10]. In these cases, individual components of the rate of strain tensor influence differently and distinctively the various terms appearing in the Reynolds stress budget, particularly the turbulence production. This mechanism cannot be captured by LEVMs since the production of turbulent kinetic energy in a Boussinesq context relies on the norm of the rate of strain tensor only.

One of the biggest advantages of DRSMs with respect to lower order closures is the exact turbulence production term, which appears in the transport equations for the Reynolds stress tensor derived directly from the Navier-Stokes equations. However, the closure problem is not automatically solved by differential models; it is only shifted, as still higher correlations of fluctuating quantities are introduced. Models have to be found for the dissipation, redistribution due to pressure-strain interaction, and turbulent diffusion of Reynolds stresses. Two DRSMs of different complexity will be evaluated in this paper, i.e. the SSG/LRR-ω and the JH-ω^h model.

The SSG/LRR-ω model has been developed specifically in view of application to complex aerodynamic flows. It is a hybrid model in which the quadratic SSG [27] and the linear LRR [17] pressure-strain models are combined. Menter's BSL ω-equation is employed for the scale determining variable along with the blending function F_1. The latter is also used to blend between the LRR model (close to solid walls) and the SSG model (away from walls). For more details on the SSG/LRR-ω model, the interested reader is referred to the original publications by Eisfeld and co-workers [3, 4]. The version used in this study is documented in [22].

On the other hand, the Jakirlić/Hanjalić model has been developed with particular attention to the exact reproduction of mean flow quantities and turbulent statistics in building block flows. For this purpose, the terms in the Reynolds stress transport equations as well as the dissipation rate equation were calibrated to reproduce the behaviour of their exact counterparts. In particular, using DNS data to compute the model terms, a system of equations was obtained and solved for the unknown model's coefficients. Analysis showed that these can be expressed as functions of turbulence anisotropy invariants and the turbulence Reynolds number [9, 12]. Originally, the model was based on the dissipation rate ϵ. Jakirlić showed that if the homogeneous dissipation rate ϵ^h is used instead, wall limits for the normalised dissipation components are satisfied automatically [13]. While adapting the model to be used in the context of scale-adaptive simulations, Maduta suggested that the specific homogeneous dissipation rate ω^h should be employed [18]. The current model is based on this latest formulation and termed JH-ω^h.

The transport equation for Reynolds stresses in the JH-ω^h model reads:

$$\frac{D\overline{\rho u_i'' u_j''}}{Dt} = \overline{\rho} P_{ij} - \overline{\rho}\epsilon_{ij}^h + \overline{\rho}\left(\Pi_{ij,1} + \Pi_{ij,2} + \Pi_{ij}^w\right) + \frac{\partial}{\partial x_k}\left[\left(\frac{1}{2}\mu + \frac{2C_S}{3C_\mu}\mu_T\right)\frac{\partial \widetilde{u_i'' u_j''}}{\partial x_k}\right] \tag{1}$$

where P_{ij} denotes the production, ϵ_{ij}^h the homogeneous dissipation and Π_{ij} the components of the pressure-strain redistribution term. A simple gradient diffusion (SGD) approach is used to close the turbulent diffusion correlations. Indices occurring twice within a product imply summation over the three spatial directions. As mentioned above, the production term

$$P_{ij} = -\left(\widetilde{u_i'' u_k''}\frac{\partial \tilde{u}_j}{\partial x_k} + \widetilde{u_j'' u_k''}\frac{\partial \tilde{u}_i}{\partial x_k}\right) \tag{2}$$

is exact. The pressure-strain correlation is traditionally split into a slow part $\Pi_{ij,1}$, a rapid part $\Pi_{ij,2}$, and a contribution due to the presence of solid walls Π_{ij}^w. Its components are given by

$$\Pi_{ij,1} = -C_1 \epsilon^h a_{ij} \tag{3}$$

$$\Pi_{ij,2} = k\left[C_3 S_{ij}^* + C_4\left(a_{ip}S_{pj} + a_{jp}S_{pi} - \frac{2}{3}a_{pq}S_{pq}\delta_{ij}\right) + C_5\left(a_{ip}W_{pj} + a_{jp}W_{pi}\right)\right] \tag{4}$$

$$\Pi_{ij}^w = C_1^w f_w \frac{\epsilon^h}{k}\left(\widetilde{u_k''u_m''}n_k n_m \delta_{ij} - \frac{3}{2}\widetilde{u_i''u_k''}n_k n_j - \frac{3}{2}\widetilde{u_k''u_j''}n_k n_i\right) \tag{5}$$
$$+ C_2^w f_w \left(\Pi_{km,2} n_k n_m \delta_{ij} - \frac{3}{2}\Pi_{ik,2} n_k n_j - \frac{3}{2}\Pi_{kj,2} n_k n_i\right).$$

All closures are formulated in terms of the mean strain rate and vorticity tensors

$$S_{ij} = \frac{1}{2}\left(\frac{\partial \tilde{u}_i}{\partial x_j} + \frac{\partial \tilde{u}_j}{\partial x_i}\right), \quad S_{ij}^* = S_{ij} - \frac{1}{3}S_{qq}\delta_{ij}, \quad W_{ij} = \frac{1}{2}\left(\frac{\partial \tilde{u}_i}{\partial x_j} - \frac{\partial \tilde{u}_j}{\partial x_i}\right), \tag{6}$$

the Reynolds stress anisotropy tensor

$$a_{ij} = \frac{\widetilde{u_i''u_j''}}{k} - \frac{2}{3}\delta_{ij} \tag{7}$$

and the wall normal vector $\mathbf{n}$. The influence of the wall is blended by a function dependent on the ratio of the turbulence length scale to the distance to the wall

$$f_w = \min\left[\frac{k^{\frac{3}{2}}}{2.5\epsilon^h y_n}, 1.4\right]. \tag{8}$$

The anisotropy of the dissipation tensor is directly coupled to the anisotropy of the Reynolds stress tensor

$$\epsilon_{ij}^h = \epsilon^h\left[\frac{2}{3}\delta_{ij} + f_s a_{ij}\right] \quad \text{with} \quad f_s = 1 - \sqrt{AE}^2. \tag{9}$$

All model coefficients are functions of invariants of a_{ij} and its dissipation counterpart $e_{ij} = f_s a_{ij}$. The second and third invariants as well as the two-component parameter are given by

$$A_2 = a_{ij}a_{ji}, \quad A_3 = a_{ij}a_{jk}a_{ki} \quad \text{and} \quad A = 1 - \frac{9}{8}(A_2 - A_3) \tag{10}$$

Table 1 Coefficients of Reynolds stress and dissipation rate equations for the JH-ω^h model

Model	Coefficient			Model	Coefficient		
$\Pi_{ij,1}$	C_1	$=$	$C + \sqrt{AE}^2$	Π_{ij}^w	C_1^w	$=$	$\max\left[1.0 - 0.7C, 0.3\right]$
	C	$=$	$2.5AF^{\frac{1}{4}} f$		C_2^w	$=$	$\min\left[A, 0.3\right]$
	F	$=$	$\min\left[0.6, A_2\right]$	Diffusion	C_S	$=$	0.082
	f	$=$	$\min\left[\left(\frac{Re_T}{150}\right)^{\frac{3}{2}}, 1\right]$	ω^h	α	$=$	0.44
	Re_T	$=$	$\frac{\bar{\rho}k^2}{\mu\epsilon^h}$		β	$=$	0.072
$\Pi_{ij,2}$	C_3	$=$	$\frac{4}{3}C_2$		σ_ω	$=$	0.9091
	C_4	$=$	C_2		σ_d	$=$	0.25
	C_5	$=$	$-C_2$		$C_{\epsilon 3}$	$=$	0.3
	C_2	$=$	$0.8\sqrt{A}$				

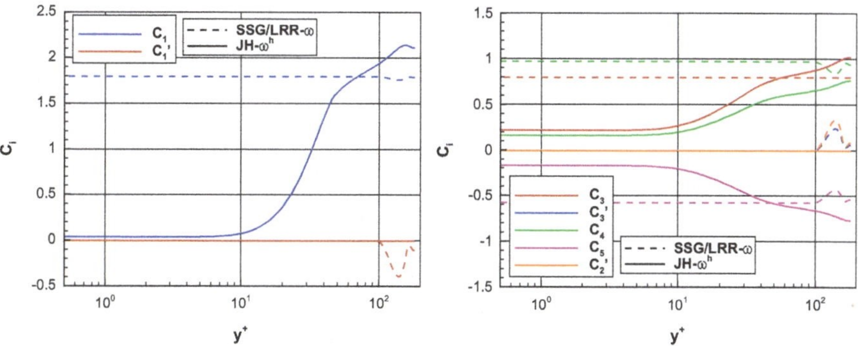

Fig. 1 Coefficient functions for slow and rapid pressure-strain terms of JH-ω^h model compared to SSG/LRR-ω model in turbulent plane channel flow

and likewise for the dissipation anisotropy e_{ij}. The coefficients of the JH-ω^h model are summarised in Table 1. In contrast to the SSG/LRR-ω model, all pressure-strain terms are tensorially linear in the anisotropy tensor. The complexity of this model lies in the variable coefficients. In the rapid part of the pressure-strain correlation $\Pi_{ij,2}$ these are chosen so that a classic isotropisation-of-production term with a variable coefficient C_2 is obtained. Furthermore, the model introduces explicit modelling of near-wall effects which Eisfeld's model does not consider.

The major difference between the two RSMs is the treatment of coefficients in the slow and rapid parts of the pressure-strain term. While they are practically constant throughout the boundary layer and only blended by Menter's F_1 function at the boundary layer edge for the SSG/LRR-ω model, they are functions of turbulence anisotropy invariants for the JH-ω^h model. The variation of the coefficient functions for the slow part (left) and the rapid part (right) is plotted in Fig. 1 using results obtained for turbulent plane channel flow. Close to solid walls, wall-normal turbulent fluctuations are damped more than the wall-parallel ones. Low Reynolds DRSMs such as the JH-ω^h model try to model this effect in different ways.

In addition to explicit wall-reflection terms, the coefficients of the slow and rapid parts of the pressure-strain redistribution are reduced depending on the two-component parameter A. No such treatment is applied by the SSG/LRR-ω model. The effect on the normal Reynolds stress components can be seen below in Fig. 2.

The homogeneous dissipation rate is related to the specific homogeneous dissipation rate by

$$\epsilon^h = C_\mu k \omega^h \tag{11}$$

with $C_\mu = 0.09$. For this quantity, a transport equation can be derived from the transport equation for ϵ^h. The employed version of the equation

$$\begin{aligned}\frac{D\left(\overline{\rho}\omega^h\right)}{Dt} =& \frac{\partial}{\partial x_i}\left[\left(\frac{1}{2}\mu + \sigma_\omega \mu_T\right)\frac{\partial \omega^h}{\partial x_i}\right] + \alpha \frac{\overline{\rho}\omega^h}{2k} P_{qq} - \beta \overline{\rho}\left(\omega^h\right)^2 \\ &+ \frac{2}{k}\left(\frac{1}{2}\mu + \sigma_d \mu_T\right)\max\left[\frac{\partial \omega^h}{\partial x_i}\frac{\partial k}{\partial x_i}, 0\right] + \frac{C_{\epsilon 3}\mu}{C_\mu}\frac{\widetilde{u_p'' u_q''}}{\epsilon^h}\frac{\partial^2 \tilde{u}_i}{\partial x_p \partial x_l}\frac{\partial^2 \tilde{u}_i}{\partial x_q \partial x_l}\end{aligned} \tag{12}$$

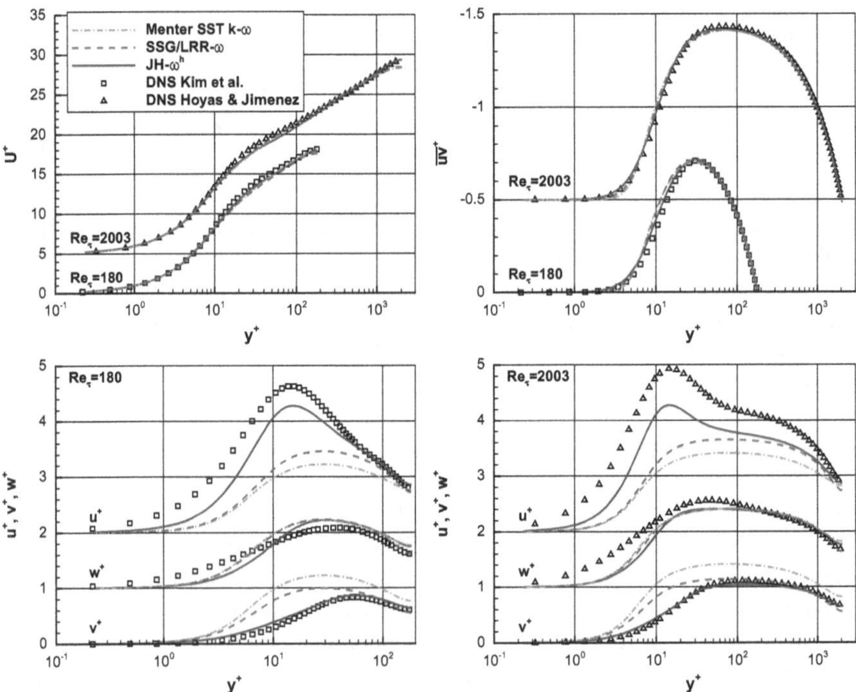

Fig. 2 Turbulent plane channel flow at $Re_\tau = 180$ [15] and $Re_\tau = 2003$ [11]. Normalised velocity profile (*top left*), shear stress (*top right*) and normal stresses (*bottom*) are compared for Menter SST k-ω, SSG/LRR-ω and JH-ω^h turbulence models

differs from the exactly transformed version in several aspects. The most important one is the limitation of the cross diffusion term to positive values in analogy to Menter's BSL equation. This improves stability in complex test cases at the expense of the accuracy of prediction of normal stresses in the viscous sublayer. Evaluation of the turbulent viscosity μ_T in a turbulent plane channel flow resulted in the formulation

$$\mu_T = 0.144 A\overline{\rho} k^{\frac{1}{2}} \max\left[10\eta_K, L\right] \quad \text{with} \quad \eta_K = \left(\frac{\nu^3}{\epsilon^h}\right)^{\frac{1}{4}} \quad \text{and} \quad L = \frac{k^{\frac{3}{2}}}{\epsilon^h}. \tag{13}$$

It is used to model the diffusion of the Reynolds stresses and the dissipation rate. In the Navier-Stokes equations it serves a mere numerical stabilisation purpose by increasing the diagonal dominance of the implicit solution matrix [22].

At solid walls the Reynolds stresses vanish as prescribed by the no-slip condition. For the dissipation rate the Taylor microscale [14] is employed to derive the following formulation for ω^h for the first cell away from the wall:

$$\omega^h\Big|_{\text{first cell}} = \frac{\nu}{C_\mu y^2}, \qquad \frac{\partial \omega^h}{\partial \mathbf{n}}\Bigg|_{\text{wall}} = 0. \tag{14}$$

The gradient is set to zero which is physically incorrect; however, this choice has no influence as long as the diffusion of ω^h is computed using only directly neighbouring cells. This treatment in the JH-ω^h model differs from the SSG/LRR-ω model, where ω is prescribed at the wall according to the suggestion by Menter [19].

3 Numerical Method

All computations were performed using the DLR flow solver for turbomachinery applications TRACE. TRACE is a hybrid grid, multi-block, compressible, implicit Navier-Stokes code based on the finite volume method. It has been developed for over 20 years at the DLR Institute of Propulsion Technology and is designed to meet the specific requirements of simulating turbomachinery flows [1]. Within the RANS framework, the turbulence transport equations are solved with a second-order accurate, conservative, segregated solution method [21]. The source terms for the Reynolds stresses and dissipation rate are linearised and treated implicitly [22].

A key to a more robust solution method was the introduction of explicit realisability constraints for all six Reynolds stress tensor components, as theoretically investigated by Schumann [26]. Since the Reynolds stress equations are solved in a segregated manner, it is possible that during the iterative solution procedure one of the normal components $\widetilde{u''_\alpha u''_\alpha}$ (no summation over Greek indices) violates the realisability condition of positive normal stresses. This limit is explicitly enforced;

however, it is also very important to satisfy the constraints on the shear stresses resulting from the Cauchy–Schwarz inequality

$$\widetilde{u''_\alpha u''_\beta}^2 \leq \widetilde{u''_\alpha u''_\alpha} \cdot \widetilde{u''_\beta u''_\beta} \tag{15}$$

as otherwise the production term of Reynolds stresses would yield unphysical values which can lead to a diverging solution. Realisability is, therefore, enforced after the update of all Reynolds stress tensor components in the following order:

1. Limit normal stresses to positive values

$$\widetilde{u''_\alpha u''_\alpha} \to \max\left[\widetilde{u''_\alpha u''_\alpha}, \epsilon\right] \tag{16}$$

   with a small positive value ϵ to avoid divisions by zero.
2. Limit shear stresses according to limited normal stresses

$$\widetilde{u''_\alpha u''_\beta} \to \operatorname{sgn}\left(\widetilde{u''_\alpha u''_\beta}\right) \sqrt{\widetilde{u''_\alpha u''_\alpha} \cdot \widetilde{u''_\beta u''_\beta}} \quad \text{if} \quad \widetilde{u''_\alpha u''_\beta}^2 > \widetilde{u''_\alpha u''_\alpha}\widetilde{u''_\beta u''_\beta}. \tag{17}$$

Another crucial factor for stability is the linearisation of the Reynolds stress source terms. Since no coupling between the Reynolds stress equations is considered, only derivatives of the source term by the respective Reynolds stress have to be calculated, i.e.

$$\frac{\partial R_{\alpha\beta}}{\partial \overline{\rho u''_\gamma u''_\delta}} = 0 \quad \text{if} \quad \alpha \neq \gamma, \beta \neq \delta. \tag{18}$$

For the production and destruction terms $\overline{\rho} P_{ij}$ and $\overline{\rho}\epsilon_{ij}$ the derivatives are straightforward:

$$\frac{\partial\left(\overline{\rho} P_{\alpha\beta}\right)}{\partial\left(\overline{\rho u''_\alpha u''_\beta}\right)} = -(S_{\alpha\alpha} + S_{\beta\beta}), \tag{19}$$

$$\frac{\partial\left(\overline{\rho}\epsilon_{\alpha\beta}\right)}{\partial\left(\overline{\rho u''_\alpha u''_\beta}\right)} = \frac{1}{3} C_\mu \omega \delta_{\alpha\beta}. \tag{20}$$

All coefficient functions were held constant in the following derivatives. For the slow pressure-strain term $\overline{\rho}\Pi_{ij,1}$, $\mu_T = \overline{\rho} k/\omega$ was held constant in the derivative to simplify the resulting expressions:

$$\begin{aligned} \frac{\partial\left(\overline{\rho}\Pi_{\alpha\beta,1}\right)}{\partial\left(\overline{\rho u''_\alpha u''_\beta}\right)} &= -C_\mu C_1 \left(1 - \frac{1}{3}\delta_{\alpha\beta}\right) \\ &+ \frac{2}{3} C_\mu C'_1 \frac{\omega}{k}\left[2k - \frac{3}{2}\left(\widetilde{u''_\alpha u''_\alpha} + \widetilde{u''_\beta u''_\beta}\right)\right]\left(1 - \frac{2}{3}\delta_{\alpha\beta}\right). \end{aligned} \tag{21}$$

The most complex expression was obtained for the Jacobian of the rapid pressure-strain term:

$$\frac{\partial(\bar{\rho}\Pi_{\alpha\beta,2})}{\partial\left(\overline{\rho u''_\alpha u''_\beta}\right)} = \frac{1}{2}C_3 S_{\alpha\beta}\delta_{\alpha\beta} + C_4\left[S_{\alpha\alpha} + S_{\beta\beta} - \frac{2}{3}\left(2S_{\alpha\beta} - \frac{1}{3}S_{qq}\right)\delta_{\alpha\beta}\right] \qquad (22)$$

$$-\frac{C'_2}{k}\left[P_k - 4\widetilde{u''_\alpha u''_\beta}S_{\alpha\beta} + \left(2k + \widetilde{u''_\alpha u''_\beta}\right)S_{\alpha\beta}\delta_{\alpha\beta}\right]\left(1 - \frac{1}{3}\delta_{\alpha\beta}\right).$$

Here, in the term including P_k, numerator and denominator were multiplied by k and k in the denominator was held constant in the derivative. According to Wilcox [29], only negative source terms are linearised while positive source terms are treated explicitly. This decision is evaluated on a cell-by-cell basis depending on the local value of the source term Jacobian. The increased diagonal dominance of the LHS matrix leads to a more stable scheme.

4 Model Validation

The implemented DRSMs were validated using a series of building block flows. In this work, the results of the simulations of the turbulent flow in a plane channel are reported as an example. Since the problem is essentially one dimensional, the case is set up as follows. Periodic boundary conditions are applied in the streamwise direction which is resolved with one cell. The spanwise direction is also resolved with one cell and symmetry boundary conditions. Only in the wall normal direction, flow quantities are expected to vary. A series of meshes with 48, 96 and 192 cells in this direction is used and mesh convergent results are reported for the finest mesh. The pressure gradient is introduced into the momentum equations by means of a volume source term whose strength is iterated until the desired Reynolds number based on friction velocity Re_τ is reached.

Figure 2 shows the results at $Re_\tau = 180$ [15] and $Re_\tau = 2003$ [11]. All quantities are made non-dimensional by the friction velocity u_τ. The velocity profile (top left) is predicted to a similar degree of accuracy by the Menter SST k-ω and the SSG/LRR-ω models. Especially the prediction of the logarithmic region at y^+ values towards the channel centre line is improved by the JH-ω^h model. While the shear stress is predicted well by all tested turbulence models with only marginal differences, the conceptual advantage of DRSMs becomes evident if normal stresses (bottom) are considered. In contrast to the LEVM, which erroneously predicts isotropic normal stresses throughout the boundary layer, both DRSMs show normal stress anisotropy to different extents. The SSG/LRR-ω model increases the streamwise normal stress by the same amount it decreases the wall normal component in the logarithmic and buffer regions. It is, however, not able to reproduce the peak in the streamwise component and the strong damping of the wall normal component as given by the DNS data. The JH-ω^h model is able to qualitatively capture the peak in u^+ and the damping of v^+ owing to the functional dependency of pressure-strain term coefficients on turbulence anisotropy invariants. On the other hand, quantitative agreement including the asymptotic behaviour towards the wall is not as well achieved as it is by the original JH-ϵ model [8].

5 Compressor Cascade Flow

The low speed compressor cascade investigated experimentally by Muthanna [23] and Tang [28] (operated at $Ma = 0.07$ and $Re = 400{,}000$) is representative of a turbomachinery flow, characterised by complex 3D flow features. The flow in the tip-gap was investigated with Laser Doppler Velocimetry [28] while a hot wire probe was used to scan the passage flow [24]. The cascade with a pitch of 236 mm was built of GE Rotor B section blades with a chord length and blade height of 254 mm staggered at an angle of 56.9° resulting in an axial chord length of $c_a = 138.68$ mm. The size of the tip-gap between the blade and the casing amounted to 1.65 % of the blade height. The inflow conditions were given by $U_\infty = 24.5$ m/s at an angle of 65.1°, a Mach number of 0.073 and a Reynolds number based on chord length of 400,000. An overview of the numerical set-up is given in Fig. 3. Due to the low Mach number, a local low Mach preconditioning of the type proposed by Turkel was employed [5]. To achieve an appropriate representation of the tip-gap flow, 34 cells were placed between the blade tip and the casing. One passage of the compressor cascade was computed on a mesh with 2.7 million cells distributed to 19 blocks and low Reynolds resolution ($y^+ < 1$) at all solid walls. Non-reflecting boundary conditions were used at inlet and outlet, and periodic boundary conditions were used

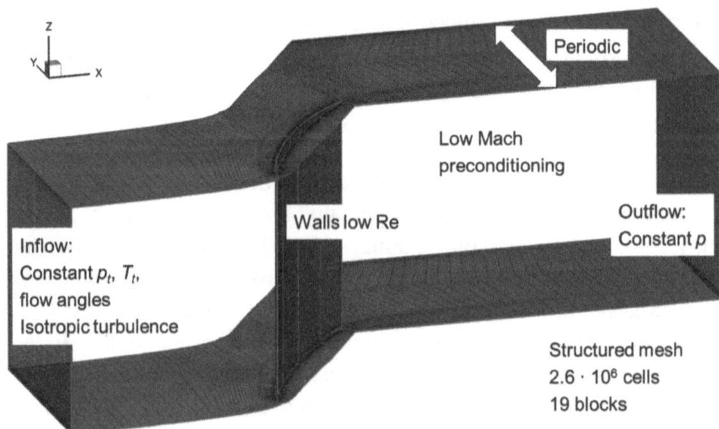

Fig. 3 Summary of numerical setup of Virginia Tech compressor cascade computation

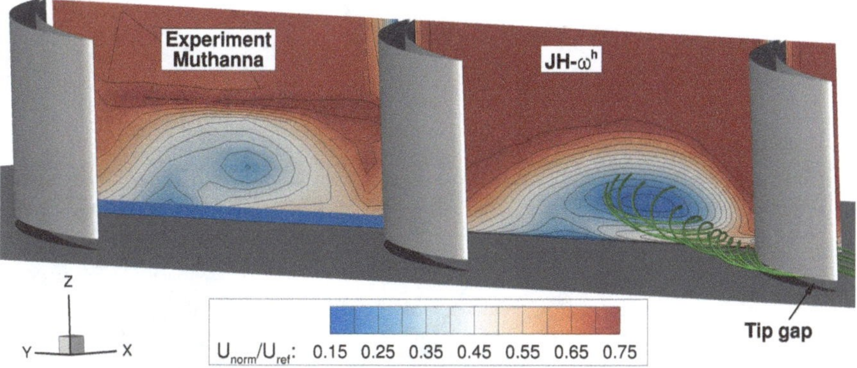

Fig. 4 Illustration of tip-leakage flow in the Virginia Tech compressor cascade. Measured mean velocity in blade passage is compared to prediction by JH-ω^h DRSM

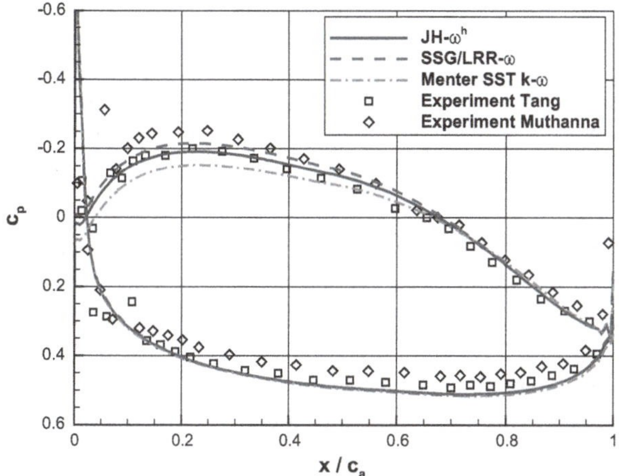

Fig. 5 Pressure coefficient c_p at midspan of Virginia Tech compressor cascade

in the pitch-wise direction. For all turbulence models tested in this work the same inflow boundary conditions, i.e. isotropic turbulence, were used.

An overview of the flow topology is shown in Fig. 4. The flow through the tip-gap of 1.65 % blade height leads to the development of the tip-gap vortex visualised by the streamlines. Qualitatively the velocity deficit in the vortex core predicted with the JH-ω^h model is compared to the measured data at $x/c_a = 0.98$, c_a being the axial chord length. The pressure distribution on the blade at midspan, predicted by the different turbulence models (Fig. 5), lies within the experimental scatter, confirming that the boundary conditions were chosen correctly.

Muthanna determined the centre of the vortex as the location of the maximal streamwise vorticity [23]. From the simulation data, the vortex core was determined using the λ_2-criterion. The position of the vortex core (spheres) at four measurement

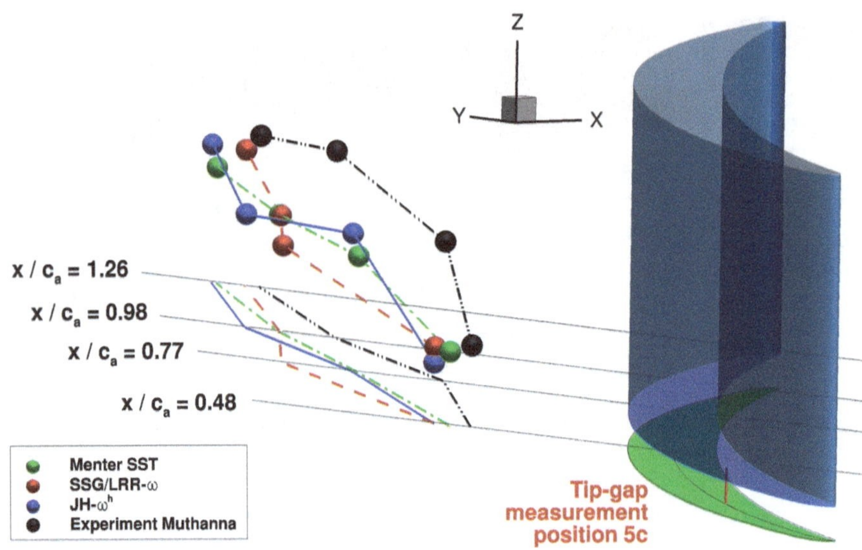

Fig. 6 Prediction of tip-gap vortex centre. Spheres represent vortex centre at different measurement planes. Its trajectory is projected to x-y plane for better comparability. The position for the examined tip-gap measurements is shown in *red*

planes is shown in Fig. 6. To facilitate comparison, its path is also projected onto the side wall (corresponding lines without symbols). From these results it can be argued that none of the employed turbulence models shows a clear advantage over the others and that all of them predict a path that is comparable to the experiment. Similar conclusions can be drawn from the prediction of the separation line on the end wall plotted in Fig. 7. The streaklines are colored with surface pressure. While differences between the models are rather subtle, some improvements can be seen from the simplest to the most complex model. Going from the Menter SST k-ω to the SSG/LRR-ω model yields an improvement in separation prediction close to the blade. In addition, the JH-ω^h model is able to move the predicted separation line towards the experiment further downstream.

Tang measured the mean velocities and Reynolds stresses at various stations in the tip-gap [28]. Due to space constraints, only the station 5c was selected to be shown in this paper as representative, with its location illustrated in Fig. 6. To obtain a representation which is independent of the selected coordinate system, invariants of the Reynolds stress anisotropy tensor a_{ij}, given by Eq. (10), are plotted in Fig. 8 instead of Reynolds stress tensor components. Close to a solid wall, turbulence is expected to tend towards the two-component limit with $A = 0$. Towards the end wall, experiments show such a trend except for the points closest to the wall, whereas towards the blade tip wall, no such trend can be observed, suggesting that the velocities were possibly not measured down to the wall. It cannot be expected from the LEVM to correctly predict turbulence anisotropy but also the high Reynolds SSG/LRR-ω DRSM is not able to capture the peak of anisotropy at the

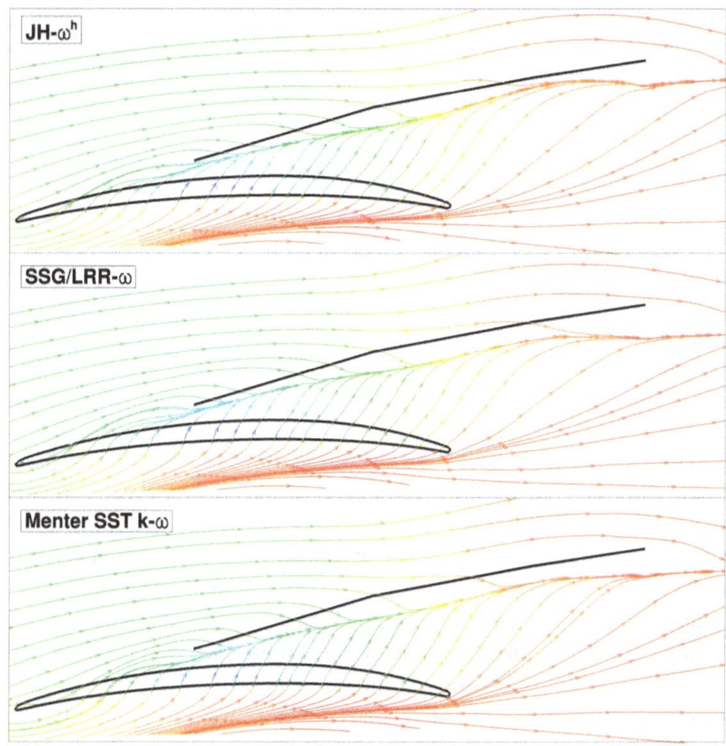

Fig. 7 Prediction of separation line on end wall. Blade profile and separation line from oil flow visualisation [23] are plotted in *black*. The *streak lines* are colored with surface pressure

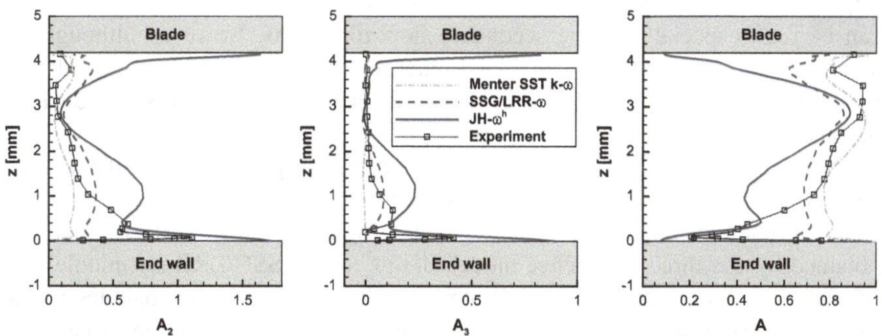

Fig. 8 Reynolds stress anisotropy invariants in tip-gap at measurement position 5c

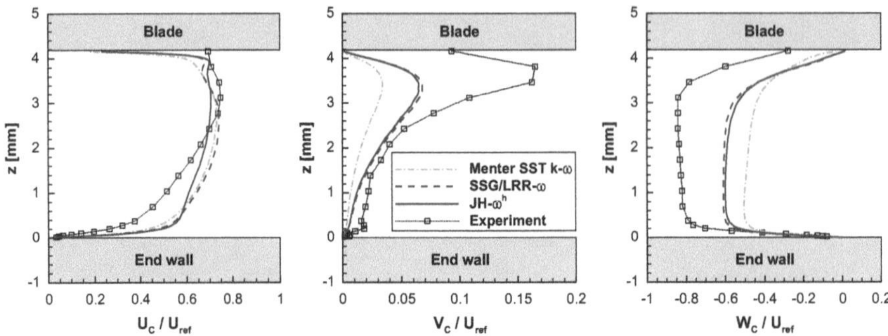

Fig. 9 Velocity components in tip-gap at measurement position 5c

end wall. The low Reynolds JH-ω^h DRSM formulation, on the contrary, predicts this peak in line with the channel flow results and displays the correct two-component turbulence behaviour at both solid walls in the tip-gap. Below the blade-tip boundary layer, turbulence reaches a nearly isotropic state with $A \to 1$, which is predicted by both DRSMs.

The mean velocity is shown in a coordinate system aligned with the chord of the blade. U_C is in the direction of the chord, V_C points in the spanwise direction and W_C is the blade-to-blade direction. Figure 9 shows the mean velocity components in the specified coordinate system normalised by the inflow velocity U_{ref}. The measured velocities do not vanish at the blade tip wall which is in line with the findings concerning the turbulence anisotropy. Whilst an almost symmetric chordwise velocity profile is predicted by the LEVM, experiments show a higher velocity near the blade tip than near the end wall. This asymmetry is predicted qualitatively by the JH-ω^h and partly by the SSG/LRR-ω DRSM. Furthermore, improvements of the results obtained with DRSMs as compared to LEVM results can be seen especially in the secondary flow directions. However, although the structure of the turbulence is predicted much better when DRSMs are employed, the improvement in quantitative agreement of mean velocity data is still far from optimal.

While vortex breakdown is one mechanism resulting in blockage of the blade passage towards off-design operating conditions, corner separation at the blade-end wall junction can also be a limiting factor. Figure 10 shows surface streaklines obtained by the three turbulence models JH-ω^h (left), SSG/LRR-ω (middle) and Menter SST k-ω (right). LEVMs notoriously predict separation bubbles whose extent in spanwise direction equals approximately their extent in streamwise direction. DRSMs, on the other hand, are able to predict asymmetric bubbles. Unfortunately, only few experiments are available documenting such flow topology features, which are of prime importance for turbomachinery performance prediction. This underlines the need for more, highly accurate measurements specifically addressing topological issues in realistic configurations.

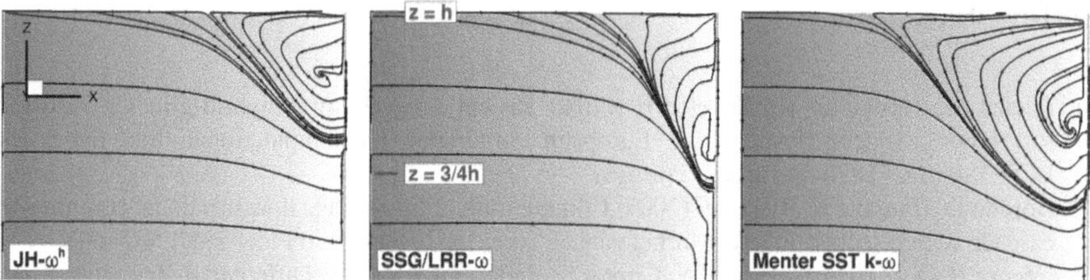

Fig. 10 Comparison of corner separation on blade wall opposite to tip-gap predicted by JH-ω^h (*left*), SSG/LRR-ω (*middle*) and Menter SST k-ω (*right*)

6 Conclusions

The flow in a low speed compressor cascade with tip-gap was simulated using DRSMs of different complexity. Whereas the SSG/LRR-ω model employs coefficients that are blended by a single empirical function, the JH-ω^h model introduces variable coefficients which are deduced from analysis of DNS data. Furthermore, the latter explicitly includes near-wall effects in the pressure-strain term and through the use of the specific homogeneous dissipation rate as scale determining variable. Improvements in the prediction of turbulence anisotropy could, therefore, be shown in the results obtained with the JH-ω^h compared to the SSG/LRR-ω DRSM. This is in agreement with the improved normal stress prediction in the turbulent plane channel flow. Improvements in mean velocity components, however, could be shown compared to the LEVM but were not as pronounced; both DRSMs showed similar behaviour.

This leads to the real dilemma of turbulence modelling. In order to gain insight into turbulence mechanisms and derive appropriate models, highly idealised flows focussing on very few isolated effects have to be studied: basically all turbulence models are calibrated using such flows. What distinguishes the various models is how accurately the flow features of building block flows can be reproduced. However, almost all flows to which the models are applied are highly complex and feature combinations of various effects. Since the governing equations are non-linear it can *per se* not be expected that a model calibrated for a number of idealised flows yields satisfying results in a complex flow. Nevertheless, this procedure seems to be the only viable way to derive and calibrate turbulence models.

In the present study, the JH-ω^h model appears to be clearly superior to the SSG/LRR-ω model in the prediction of the building block flow. Yet, this advantage cannot be recognised as pronounced in a complex 3D flow. This raises the question of the appropriate level of complexity of DRSMs for practical simulations of such flows. It can only be answered by means of further investigations, such as the analysis of turbomachinery components including rotating frames of reference, planned by the authors.

References

1. Becker K, Heitkamp K, Kügeler E (2010) Recent progress in a hybrid-grid CFD solver for turbomachinery flows. In: V European conference on computational fluid dynamics ECCOMAS CFD 2010, Lisbon, Portugal
2. Borello D, Hanjalić K, Rispoli F (2007) Computation of tip-leakage flow in a linear compressor cascade with a second-moment turbulence closure. Int J Heat Fluid Flow 28(4):587–601
3. Cécora R-D, Eisfeld B, Probst A, Crippa S, Radespiel R (2012) Differential Reynolds stress modeling for aeronautics. AIAA-paper 2012-0465, 50th AIAA aerospace sciences meeting, Nashville, TN, USA
4. Eisfeld B, Brodersen O (2005) Advanced turbulence modelling and stress analysis for the DLR-F6 configuration. AIAA-paper 2005-4727, 23rd AIAA applied aerodynamics conference, Toronto, Canada
5. Fiedler J, di Mare F (2012) Generalised implementation of low-mach preconditioning for arbitrary three-dimensional geometries. In: 6th European congress on computational methods in applied sciences and engineering (ECCOMAS 2012), Vienna, Austria
6. Gerolymos GA, Vallet I (2007) Robust implicit multigrid Reynolds-stress model computation of 3D turbomachinery flows. J Fluids Eng 129(9):1212–1227
7. Gerolymos GA, Neubauer J, Sharma VC, Vallet I (2002) Improved prediction of turbomachinery flows using near-wall Reynolds-stress model. J Turbomach 124(1):86–99
8. Hanjalić K, Jakirlić S (1993) A model of stress dissipation in second-moment closures. Appl Sci Res 51:513–518
9. Hanjalić K, Jakirlić S (1998) Contribution towards the second-moment closure modelling of separating turbulent flows. Comput Fluids 27(2):137–156
10. Hanjalić K, Jakirlić S (2002) Second-moment turbulence closure modelling. In: Launder BE, Sandham ND (eds) Closure strategies for turbulent and transitional flows. Cambridge University Press, Cambridge, pp 47–101
11. Hoyas S, Jimenez J (2006) Scaling of the velocity fluctuations in turbulent channels up to $Re_\tau = 2003$. Physics of Fluids 18(1):011702
12. Jakirlić S (2004) A DNS-based scrutiny of RANS approaches and their potential for predicting turbulent flows. Habilitation, TU Darmstadt
13. Jakirlić S, Hanjalić K (2002) A new approach to modelling near-wall turbulence energy and stress dissipation. J Fluid Mech 459:139–166
14. Jakirlić S, Jovanović J (2010) On unified boundary conditions for improved predictions of near-wall turbulence. J Fluid Mech 656:530–539
15. Kim J, Moin P, Moser R (1987) Turbulence statistics in fully developed channel flow at low Reynolds number. J Fluid Mech 177:133–166
16. Langston LS (2001) Secondary flows in axial turbines–a review. Ann N Y Acad Sci 934:11–26
17. Launder BE, Reece G, Rodi W (1975) Progress in the development of a Reynolds-stress turbulence closure. J Fluid Mech 68:537–566
18. Maduta R, Jakirlić S (2010) Scrutinizing scale-supplying equation towards an instability sensitive second-moment closure model. In: 8th international ERCOFTAC symposium on engineering turbulence modelling and measurements - ETMM8, Marseille, France
19. Menter F (1992) Improved two-equation k-ω turbulence models for aerodynamic flows. NASA technical memorandum 103975, Moffett Field, CA, USA
20. Menter F, Kuntz M, Langtry R (2003) Ten years of industrial experience with the SST model. In: Hanjalić K, Nagano Y, Tummers M (eds) Turbulence, heat and mass transfer 4. Begell House, Inc.:pp 625–632
21. Morsbach C, di Mare F (2012) Conservative segregated solution method for turbulence model equations in compressible flows. In: 6th European congress on computational methods in applied sciences and engineering (ECCOMAS 2012), Vienna, Austria

22. Morsbach C, Franke M, di Mare F (2012) Towards the application of Reynolds stress transport models to 3D turbomachinery flows. In: 7th international symposium on turbulence, heat and mass transfer, Palermo, Sicily, Italy
23. Muthanna C (2002) The effects of free stream turbulence on the flow field through a compressor cascade. Dissertation, Virginia Polytechnic Institute and State University
24. Muthanna C, Devenport WJ (2004) Wake of a compressor cascade with tip gap, Part 1: Mean flow and turbulence structure. AIAA J 42(11):2320–2331
25. Rautaheimo PP, Salminen EJ, Sikonen TL (2003) Numerical simulation of the flow in the NASA low-speed centrifugal compressor. Int J Turbo Jet Engines 20:155–170
26. Schumann U (1977) Realizability of Reynolds-stress turbulence models. Phys Fluids 20(5):721–725
27. Speziale CG, Sarkar S, Gatski TB (1991) Modelling the pressure-strain correlation of turbulence: an invariant dynamical systems approach. J Fluid Mech 227:245–272
28. Tang G (2004) Measurements of the tip-gap turbulent flow structure in a low-speed compressor cascade. Dissertation, Virginia Polytechnic Institute and State University
29. Wilcox DC (2006) Turbulence modeling for CFD, 3rd edn. DCW Industries, La Cañada

Application of Reynolds Stress Models to Separated Aerodynamic Flows

Christopher L. Rumsey

Abstract Several variations of ω-based second-moment Reynolds stress models (RSMs) are applied to two-dimensional and three-dimensional separated aerodynamic flows. In many of these flows, widely used one- and two-equation linear eddy-viscosity turbulence models are known to be inadequate for predicting separated flow characteristics. As potentially important non-linear behavior is naturally included in RSMs, it was hoped that they might improve the separated flow predictions. However, the RSMs perform no better than the simpler models for these particular flows. Like the simpler models, the RSMs predict too little turbulence in the separated shear layer of the two-dimensional flow over a hump, which is indicative of modeling deficiencies for this class of flows. Nonetheless, the best RSM version tested offers a convenient framework for possible future model improvements.

1 Introduction

The Reynolds-averaged Navier–Stokes equations include an un-closed Reynolds stress term that must be modeled. Various closure models have been developed over the years, with different levels of fidelity. Second-moment Reynolds stress models (RSMs) represent a high level of closure, with six equations solved for the Reynolds stress tensor along with one equation for a scale-determining variable (such as turbulence energy-dissipation rate ε or specific dissipation rate ω). RSMs have been around for some time; see, for example, Hanjalić and Launder [8].

However, over the last 40 years the benefits of RSMs (which naturally include rigorous handling of stress anisotropies, streamline curvature, etc.) have tended to be outweighed by their deficiencies (more costly and stiffer equation set than one- and two-equation models with more constants to calibrate). In addition, despite the more complete modeling of the physics, RSMs generally have *not* proven to be

C.L. Rumsey (✉)
NASA Langley Research Center, Computational AeroSciences Branch, Mail Stop 128, Hampton, VA 23681, USA
e-mail: c.l.rumsey@nasa.gov

B. Eisfeld (ed.), *Differential Reynolds Stress Modeling for Separating Flows in Industrial Aerodynamics*, Springer Tracts in Mechanical Engineering,
DOI 10.1007/978-3-319-15639-2_2

consistently better than simpler models for many aerodynamic flows of interest. The reasons for this are not fully known. Partly, it may be that the simpler one- and two-equation models have been calibrated specifically for thin-shear-type aerodynamic flows, so they typically perform well, by design, for many cases of interest. When RSMs performed consistently no better (in general) for these problems, there was little incentive to continue to use them. Furthermore, most RSMs in the past have used an ε-equation for their scale-determining variable. The ε-equation tends to be less robust than the ω-equation, particularly for wall-bounded flows in adverse pressure gradient [31]. Therefore, the poor robustness of ε-based RSMs may have predisposed users in the aerodynamics community to employ the easier-to-use one- and two-equation turbulence models instead.

In this paper, we investigate the capabilities of three ω-based RSMs for several separated flows. The WilcoxRSM-w2006 model is the latest version of the stress-omega model of Wilcox [31]. It uses the LRR model [11] for its pressure-strain term. The more recently developed SSG/LRR-RSM-w2012 model [2, 4] employs a pressure-strain term that is a blend between the LRR model and the SSG model [28]. We also test a slight variant of SSG/LRR-RSM-w2012 that uses a simple diffusion model [25] rather than the generalized gradient-diffusion model [3]. In all cases, RSM results will be compared to experiment, as well as to results from the widely used Spalart–Allmaras (SA) one-equation model [27] and Menter shear-stress transport (SST) two-equation model [13]. CFD results from a less widely used two-equation explicit algebraic stress model EASMko2003-S [21] are also included for comparison. Through this assessment, strengths and weaknesses of the ω-based RSMs for separated flows will be highlighted.

2 Turbulence Model Descriptions

All turbulence models employed in this study are completely described in Rumsey [20]. However, because this paper focuses on the RSMs, most of the details for those models are provided here. The WilcoxRSM-w2006 model [31] solves for the six Reynolds stress equations and length scale equation using:

$$\frac{\partial \bar{\rho} R'_{ij}}{\partial t} + \frac{\partial (\bar{\rho} \hat{u}_k R'_{ij})}{\partial x_k} = -\bar{\rho} P_{ij} - \bar{\rho} \Pi_{ij} + \frac{2}{3} \beta^* \bar{\rho} \omega k \delta_{ij} + \frac{\partial}{\partial x_k} \left[\left(\bar{\mu} + \sigma^* \mu_t \right) \frac{\partial R'_{ij}}{\partial x_k} \right] \tag{1}$$

$$\frac{\partial (\bar{\rho} \omega)}{\partial t} + \frac{\partial (\bar{\rho} \hat{u}_k \omega)}{\partial x_k} = \frac{\alpha \bar{\rho} \omega}{k} R'_{ij} \frac{\partial \hat{u}_i}{\partial x_j} - \beta \bar{\rho} \omega^2 + \frac{\partial}{\partial x_k} \left[\left(\bar{\mu} + \sigma \mu_t \right) \frac{\partial \omega}{\partial x_k} \right] + \sigma_d \frac{\bar{\rho}}{\omega} \frac{\partial k}{\partial x_j} \frac{\partial \omega}{\partial x_j} \tag{2}$$

where $\overline{\rho} R'_{ij} \equiv \tau_{ij} = -\overline{\rho u''_i u''_j}$. Here ρ is density, u_i represents the velocity vector, and τ_{ij} is the turbulent Reynolds stress tensor. The overbar indicates a conventional time-average mean and the double-prime represents a turbulent fluctuating quantity. The quantity $\hat{u}$ in Eqs. (1) and (2) represents the Favre (density-weighted) average of velocity. The production term is $\overline{\rho} P_{ij} = \overline{\rho} R'_{ik} \frac{\partial \hat{u}_j}{\partial x_k} + \overline{\rho} R'_{jk} \frac{\partial \hat{u}_i}{\partial x_k}$, and $k = -R'_{ii}/2$ and $\mu_t = \overline{\rho} k/\omega$. The pressure-strain term is modeled via:

$$\overline{\rho}\Pi_{ij} = -C_1 \overline{\rho} \varepsilon a_{ij} + (\hat{\alpha} + \hat{\beta}) \overline{\rho} k \left(a_{ik} S_{jk} + a_{jk} S_{ik} - \frac{2}{3} a_{kl} S_{kl} \delta_{ij} \right)$$
$$+ (\hat{\alpha} - \hat{\beta}) \overline{\rho} k \left(a_{ik} W_{jk} + a_{jk} W_{ik} \right) + \left[\frac{4}{3} \left(\hat{\alpha} + \hat{\beta} \right) - \hat{\gamma} \right] \overline{\rho} k \left(S_{ij} - \frac{1}{3} S_{kk} \delta_{ij} \right) \tag{3}$$

with $a_{ij} = -\frac{R'_{ij}}{k} - \frac{2}{3}\delta_{ij}$, $\varepsilon = \beta^* k \omega$, $S_{ij} = \frac{1}{2}\left(\frac{\partial \hat{u}_i}{\partial x_j} + \frac{\partial \hat{u}_j}{\partial x_i} \right)$, and $W_{ij} = \frac{1}{2}\left(\frac{\partial \hat{u}_i}{\partial x_j} - \frac{\partial \hat{u}_j}{\partial x_i} \right)$. Closure coefficients are given by: $\hat{\alpha} = (8 + C_2)/11$, $\hat{\beta} = (8C_2 - 2)/11$, $\hat{\gamma} = (60C_2 - 4)/55$, $C_1 = 9/5$, $C_2 = 10/19$, $\alpha = 13/25$, $\beta = \beta_0 f_\beta$, $\beta^* = 9/100$, $\sigma = 0.5$, $\sigma^* = 0.6$, and $\beta_0 = 0.0708$, where $f_\beta = \frac{1 + 85\chi_\omega}{1 + 100\chi_\omega}$, $\chi_\omega = \left| \frac{W_{ij} W_{jk} \hat{S}_{ki}}{(\beta^* \omega)^3} \right|$, and $\hat{S}_{ki} = S_{ki} - \frac{1}{2}\frac{\partial \hat{u}_m}{\partial x_m}\delta_{ki}$.

The SSG/LRR-RSM-w2012 model [2, 4] solves for the six Reynolds stress equations and the length scale equation using:

$$\frac{\partial \overline{\rho} \hat{R}_{ij}}{\partial t} + \frac{\partial (\overline{\rho} \hat{u}_k \hat{R}_{ij})}{\partial x_k} = \overline{\rho} P_{ij} + \overline{\rho} \Pi_{ij} - \overline{\rho} \varepsilon_{ij} + \overline{\rho} D_{ij} \tag{4}$$

$$\frac{\partial (\overline{\rho}\omega)}{\partial t} + \frac{\partial (\overline{\rho} \hat{u}_k \omega)}{\partial x_k} = -\frac{\alpha_\omega \overline{\rho} \omega}{\hat{k}} \hat{R}_{ij} \frac{\partial \hat{u}_i}{\partial x_j} - \beta_\omega \overline{\rho} \omega^2 + \frac{\partial}{\partial x_k}\left[\left(\overline{\mu} + \sigma_\omega \frac{\overline{\rho} \hat{k}}{\omega} \right) \frac{\partial \omega}{\partial x_k} \right]$$
$$+ \sigma_d \frac{\overline{\rho}}{\omega} \max\left(\frac{\partial \hat{k}}{\partial x_j} \frac{\partial \omega}{\partial x_j}, 0 \right) \tag{5}$$

where $\overline{\rho} \hat{R}_{ij} \equiv -\tau_{ij} = \overline{\rho u''_i u''_j}$ (note that in the current notation the $\hat{R}_{ij}$ used by SSG/LRR-RSM-w2012 is the negative of the R'_{ij} used by WilcoxRSM-w2006; this is simply a matter of choice and is done to remain consistent with the original references). Here the production term is $\overline{\rho} P_{ij} = -\overline{\rho} \hat{R}_{ik} \frac{\partial \hat{u}_j}{\partial x_k} - \overline{\rho} \hat{R}_{jk} \frac{\partial \hat{u}_i}{\partial x_k}$, and $\hat{k} = \hat{R}_{ii}/2$.

The dissipation is modeled via: $\overline{\rho}\varepsilon_{ij} = \frac{2}{3}\overline{\rho}\varepsilon\delta_{ij}$, where $\varepsilon = C_\mu \hat{k}\omega$ and $C_\mu = 0.09$. The pressure-strain term is given by:

$$\overline{\rho}\Pi_{ij} = -\left(C_1\overline{\rho}\varepsilon + \frac{1}{2}C_1^*\overline{\rho}P_{kk}\right)\hat{a}_{ij} + C_2\overline{\rho}\varepsilon\left(\hat{a}_{ik}\hat{a}_{kj} - \frac{1}{3}\hat{a}_{kl}\hat{a}_{kl}\delta_{ij}\right)$$
$$+\left(C_3 - C_3^*\sqrt{\hat{a}_{kl}\hat{a}_{kl}}\right)\overline{\rho}\hat{k}S_{ij}^* + C_4\overline{\rho}\hat{k}\left(\hat{a}_{ik}S_{jk} + \hat{a}_{jk}S_{ik} - \frac{2}{3}\hat{a}_{kl}S_{kl}\delta_{ij}\right)$$
$$+C_5\overline{\rho}\hat{k}\left(\hat{a}_{ik}W_{jk} + \hat{a}_{jk}W_{ik}\right) \tag{6}$$

with $\hat{a}_{ij} = \frac{\hat{R}_{ij}}{\hat{k}} - \frac{2}{3}\delta_{ij}$ and $S_{ij}^* = S_{ij} - \frac{1}{3}S_{kk}\delta_{ij}$. All of the closure coefficients are blended using the F_1 parameter of Menter [13], which is a function of the distance to the nearest wall. The inner coefficients are: $\alpha_\omega^{(\omega)} = 0.5556$, $\beta_\omega^{(\omega)} = 0.075$, $\sigma_\omega^{(\omega)} = 0.5$, $\sigma_d^{(\omega)} = 0$, $C_1^{(\omega)} = 1.8$, $C_1^{*(\omega)} = 0$, $C_2^{(\omega)} = 0$, $C_3^{(\omega)} = 0.8$, $C_3^{*(\omega)} = 0$, $C_4^{(\omega)} = 0.5(18C_2^{(\mathrm{LRR})} + 12)/11$, $C_5^{(\omega)} = 0.5(-14C_2^{(\mathrm{LRR})} + 20)/11$, $D^{(\omega)} = 0.75C_\mu$, and $C_2^{(\mathrm{LRR})} = 0.52$. The outer coefficients are: $\alpha_\omega^{(\varepsilon)} = 0.44$, $\beta_\omega^{(\varepsilon)} = 0.0828$, $\sigma_\omega^{(\varepsilon)} = 0.856$, $\sigma_d^{(\varepsilon)} = 1.712$, $C_1^{(\varepsilon)} = 1.7$, $C_1^{*(\varepsilon)} = 0.9$, $C_2^{(\varepsilon)} = 1.05$, $C_3^{(\varepsilon)} = 0.8$, $C_3^{*(\varepsilon)} = 0.65$, $C_4^{(\varepsilon)} = 0.625$, $C_5^{(\varepsilon)} = 0.2$, and $D^{(\varepsilon)} = 0.22$.

The generalized gradient-diffusion model of Daly and Harlow [3] is used in the SSG/LRR-RSM-w2012 model:

$$\overline{\rho}D_{ij} = \frac{\partial}{\partial x_k}\left[\left(\overline{\mu}\delta_{kl} + D\frac{\overline{\rho}\hat{k}\hat{R}_{kl}}{\varepsilon}\right)\frac{\partial \hat{R}_{ij}}{\partial x_l}\right] = \frac{\partial}{\partial x_k}\left[\left(\overline{\mu}\delta_{kl} + D\frac{\overline{\rho}\hat{R}_{kl}}{C_\mu\omega}\right)\frac{\partial \hat{R}_{ij}}{\partial x_l}\right] \tag{7}$$

However, for some cases the generalized gradient-diffusion model has been found to cause numerical problems. Therefore, a simplified variant has also been developed, termed SSG/LRR-RSM-w2012-SD. Here, "SD" stands for simple diffusion, which is modeled via:

$$\overline{\rho}D_{ij} = \frac{\partial}{\partial x_k}\left[\left(\overline{\mu} + D\frac{\overline{\rho}\hat{k}^2}{\varepsilon}\right)\frac{\partial \hat{R}_{ij}}{\partial x_k}\right] = \frac{\partial}{\partial x_k}\left[\left(\overline{\mu} + D\frac{\overline{\rho}\hat{k}}{C_\mu\omega}\right)\frac{\partial \hat{R}_{ij}}{\partial x_k}\right] \tag{8}$$

with $D = 0.5C_\mu F_1 + \frac{2}{3}0.22(1 - F_1)$.

3 Numerical Method

The above-mentioned RSMs have been implemented in two different NASA Langley CFD codes: CFL3D [10, 19] and FUN3D [1, 5]. However, for this paper only simulations using CFL3D have been performed.

CFL3D is a cell-centered, structured, multiblock, multigrid Reynolds-averaged Navier–Stokes (RANS) code. The mean-flow convective terms are discretized with third-order upwind-biased spatial differencing, and viscous terms are discretized with second-order central differencing. The flux-difference splitting method of Roe [16] is employed to obtain fluxes at the cell faces. Advancement in time is accomplished via backward Euler, with an implicit approximate factorization scheme. The turbulence equations are solved de-coupled from the mean flow; they are also solved with the approximate factorization method. The advection terms in the turbulence equations are solved with first-order upwind discretization. Destruction and diffusion terms in the turbulence models are treated implicitly, and production terms are treated explicitly.

For the RSMs in CFL3D, wall boundary conditions are zero for the Reynolds stresses, and ω_{wall} is set according to the method of Menter [13]. At far field boundaries, the normal components of the specific Reynolds stress tensor are set to $6 \times 10^{-9} a_\infty^2$, with all other components zero. The far field ω is set to $1 \times 10^{-6} \rho_\infty a_\infty^2 / \mu_\infty$.

4 Verification and Validation Studies

Validation studies have been conducted using the WilcoxRSM-w2006 model in CFL3D and FUN3D on a variety of 2-D and 3-D flows [15, 29]. Also, although not published, a limited number of internal verification studies have also been conducted using code-to-code comparisons on simple cases [20] using this model.

Verification studies have been conducted for the SSG/LRR-RSM-w2012 and SSG/LRR-RSM-w2012-SD models. In the absence of more rigorous tests such as manufactured or analytic solutions [14], we have performed studies comparing two independently coded CFD codes (CFL3D and TAU [6, 24]) applied to the same simple problems. By making use of thorough grid convergence studies in these cases, we can demonstrate that the two implementations converge to essentially the same result as the grid is refined. While not foolproof, this method can provide a great deal of confidence that the turbulence models have been implemented correctly in both codes. It is unlikely that both CFD codes would have implemented the same mistake(s). This method obviously does not account for coding mistakes that make very little difference, or that do not show up for the particular cases being tested.

Figure 1 shows results from a grid convergence study for a 2-D turbulent flat plate of length $L = 2$ and $Re_L = 5$ million. This case is described in great detail on the Turbulence Modeling Resource website [20]. A family of five successively finer structured grids were employed, ranging from 35×25 to 545×365. In Fig. 1a, plate drag coefficients from CFL3D and TAU are shown to approach nearly the same values as $h \rightarrow 0$. The SSG/LRR-RSM-w2012-SD version yielded slightly higher C_D than SSG/LRR-RSM-w2012, and the codes are consistent in this regard. In Fig. 1b, plate surface skin friction coefficient at location $x/L = 0.97$ is also demonstrated to be consistent between the two codes as $h \rightarrow 0$. Although not

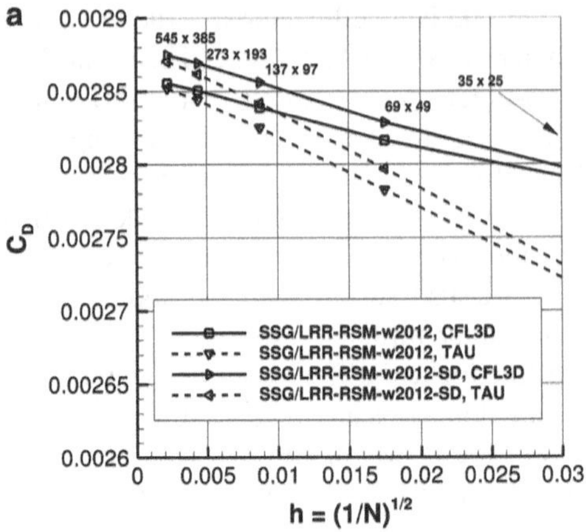

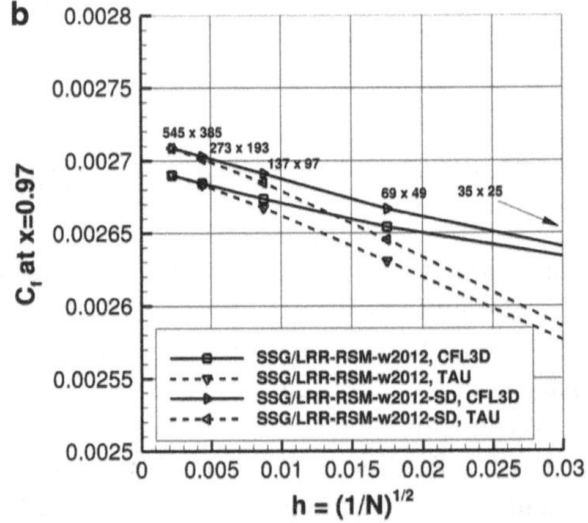

Fig. 1 Grid convergence of SSG/LRR-RSM-w2012 and SSG/LRR-RSM-w2012-SD turbulence models in two CFD codes for turbulent flat plate study (length of plate $L = 2$ units), $M = 0.2$, $Re_L = 5$ million. (**a**) Drag coefficient. (**b**) Skin friction coefficient at $x = 0.97L$

shown, both codes yielded nearly identical C_f over the entire plate when run on the finer grids. The Turbulence Modeling Resource website provides additional details for this case, showing (for example) ω and all R'_{ij} to be essentially identical between the two codes as the grid is refined. The website also provides verification results for a 2-D bump case using these two RSMs.

5 Results

Three separated-flow cases are considered. The first is a 2-D case that has been the subject of several workshops in the past, for which RANS models perform poorly in and downstream of the separation bubble [17]. The second is a 3-D transonic wing case, which was also investigated using RSMs in Cecora et al. [2]. The third is a 3-D wing-body configuration with a wing-root separation bubble.

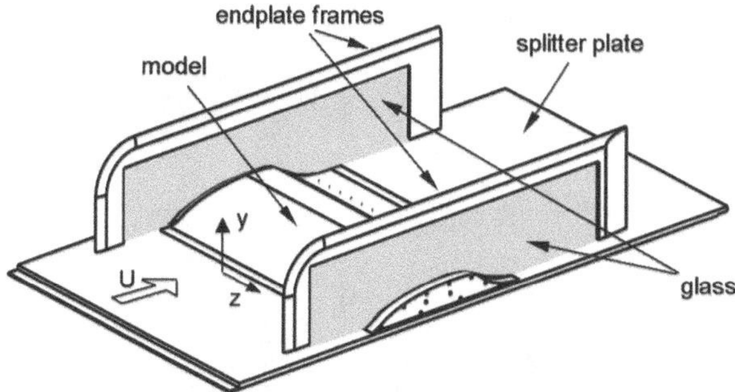

Fig. 2 Sketch of the NASA wall-mounted hump configuration

5.1 NASA Wall-Mounted Hump

The NASA wall-mounted hump case was used at a 2004 workshop [22]. The case was intended to investigate flow control, but it also included a baseline case with no blowing or suction. Because the RANS models performed poorly even for the baseline case, we repeat those computations here using RSMs. Details about the NASA hump case can also be found on a website [18].

The 2-D hump configuration is shown in Fig. 2. The CFD fine grid consisted of approximately 208,000 cells. Grid sensitivity was also explored using a medium grid consisting of every other grid point in each coordinate direction, or approximately 52,000 cells. The upstream grid extended to $x/c = -6.39$, where c represents the hump "chord." This location was found in previous RANS studies to be long enough to allow for natural development of the boundary layer to approximately match experimental upstream thickness. The downstream grid extended to $x/c = 4$. The upper boundary at $x/c = 0.90905$ corresponded to the upper tunnel wall location, solved here as a slip wall. The upper boundary was contoured to approximately account for side plate blockage effects [18].

Figure 3 shows computed bubble size using six different turbulence models. As expected based on the earlier workshop results, the SA, SST, and EASMko2003-S models all over-predicted the bubble size (reattachment location was too far downstream). Considering that the bubble length in the experiment was approximately $x_b/c = 0.445$, the SA and SST models overpredicted the length by about 35 %, and EASMko2003-S over-predicted by about 45 %. The new RSM results (all three models) also predicted the downstream reattachment location too far downstream. WilcoxRSM-w2006 predicted a bubble length larger than SA or SST, but shorter than EASMko2003-S. Both SSG/LRR-RSM-w2012 and SSG/LRR-RSM-w2012-SD predicted the smallest bubbles, but they were still too long by as much as 25 %. It should be noted that all three RSMs exhibited an unrealistic back bending of the streamline near reattachment. This is a well-known phenomenon described in Hanjalić and Jakirlić [7], due to an excessive growth of the length scale in this region. They subsequently formulated an additional term in the ε-equation to

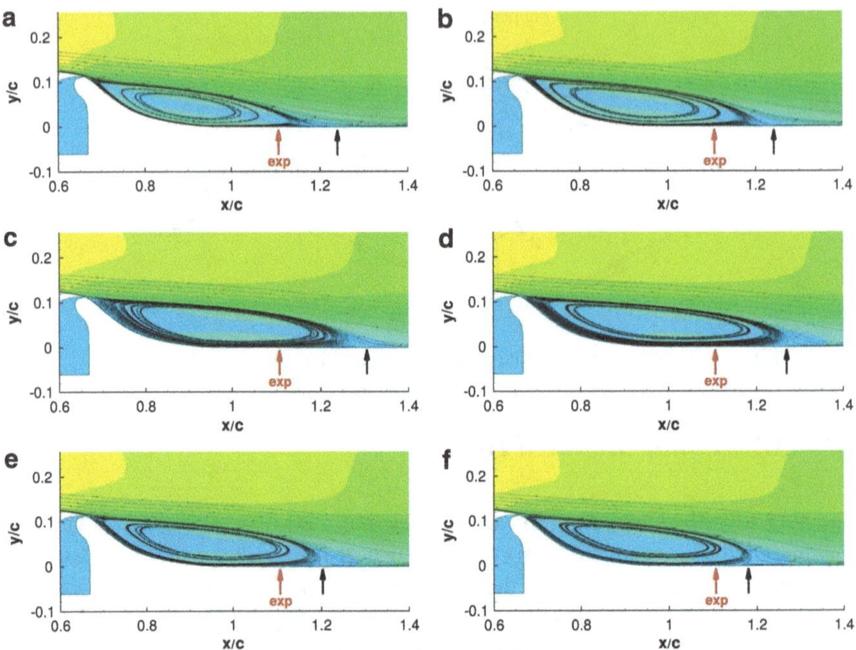

Fig. 3 Streamlines and Mach number contours of flow over baseline NASA hump model; computed reattachment point (*black arrow*) shown compared to experimental reattachment point (*red arrow*). (**a**) SA model. (**b**) SST model. (**c**) EASMko2003-S model. (**d**) WilcoxRSM-w2006 model. (**e**) SSG/LRR-RSM-w2012 model. (**f**) SSG/LRR-RSM-w2012-SD model

compensate for it. A similar fix has not yet been attempted for the ω-based models detailed here.

It is known that the main reason why the RANS models predict a bubble of excessive length is the fact that they severely under-predict the mean turbulence activity in the separated shear layer [17, 22]. Typical turbulent shear stress results at $x/c = 0.8$ are shown in Fig. 4. Here, the influence of grid size is also shown; the fine grid yielded larger peak turbulent shear stress (in magnitude) than the medium grid by a small amount. Although not shown, the effect of grid size on mean flow variables was relatively insignificant. As can be seen in the figure, all models under-predicted the peak magnitude of the turbulent shear stress by about a factor of two. The SSG/LRR-RSM-w2012 and SSG/LRR-RSM-w2012-SD predicted larger levels (in magnitude) than WilcoxRSM-w2006, which is consistent with the respective bubble sizes.

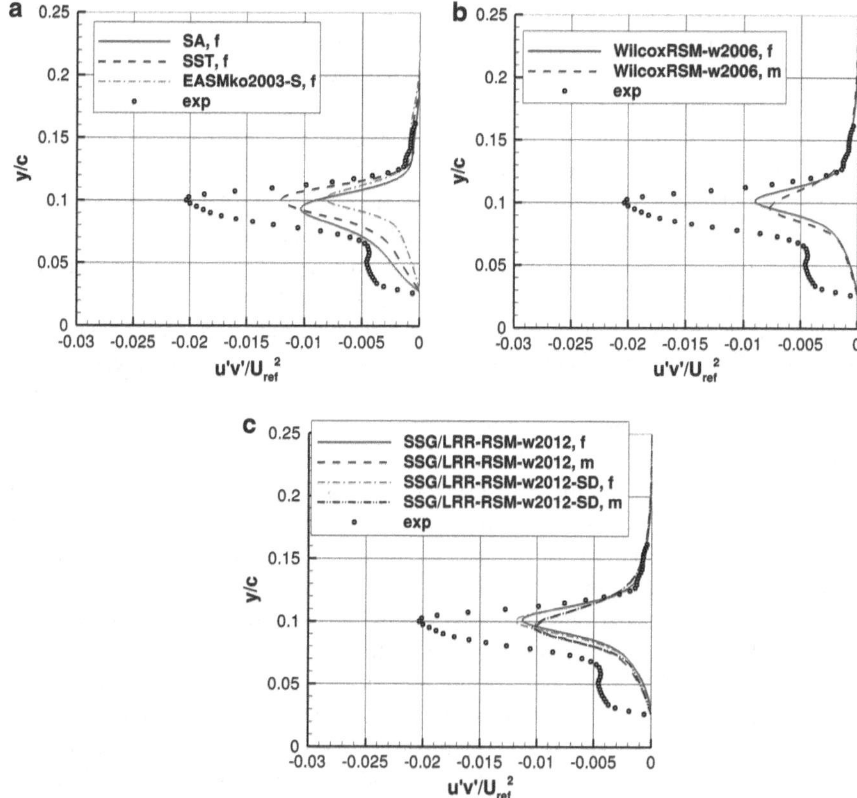

Fig. 4 Turbulent shear stress profiles at $x/c = 0.8$ for baseline NASA hump model; *f* indicates fine grid, *m* indicates medium grid. (**a**) SA, SST, and EASMko2003-S models. (**b**) WilcoxRSM-w2006 model. (**c**) SSG/LRR-RSM-w2012 and SSG/LRR-RSM-w2012-SD models

5.2 ONERA M6 Wing

The ONERA M6 wing experiments from Schmitt and Charpin [23] have been widely used over the years for validation of CFD codes and turbulence models. These cases are at transonic Mach numbers, and include conditions with upper surface shocks as well as some shock-induced separated regions. However, it is important to note that the experiment was a semi-span test that included a wall stand-off with splitter plate to reduce the influence of the wall boundary layer. Most CFD codes compute the cases using symmetry boundary conditions at the wing root, which likely introduces some error when comparing to experiment, particularly over the inner portion of the wing. This same approximation was used here.

The semi-span length of the ONERA M6 wing (not including rounded tip) was 1.1963 m. The most widely computed case has been at the conditions of $\alpha = 3.06°$, $M = 0.84$, $Re_{MAC} = 11.72 \times 10^6$, where the mean aerodynamic chord (MAC) is 0.64607 m. Although not shown, this same low angle-of-attack case was also computed for this study using SA, SST, EASMko2003-S, WilcoxRSM-w2006, and SSG/LRR-RSM-w2012. Surface pressure coefficient results from all the models were nearly identical, in reasonably good agreement with experiment and with other previously published CFD results. Instead, here we focus on the more challenging angle-of-attack of $\alpha = 4.08°$. This same condition was employed in the earlier study by Cecora et al. [2]. In that study, they showed that the SA and SST models predict a shock location that was too far forward, with massive separation over the outer half of the wing. On the other hand, two RSMs produced reasonably good results, with the SSG/LRR-RSM-w2012 model somewhat better overall than a modified form of a model from Jakirlić and Hanjalić [9].

Computations were performed on a series of grids ranging from 0.9 million cells ($289 \times 65 \times 49$) to 24 million cells ($865 \times 193 \times 145$). The outer grid extent was approximately 12 MAC from the wing. Results for the $\alpha = 4.08°$ case are shown in Fig. 5 on the finest grid. In these computations, only the WilcoxRSM-w2006 model showed signs of shock moving too far forward (near the tip), along with excessive separation. All other models—including SA, SST, and EASMko2003-S—yielded very similar results, in contrast to the results in Cecora et al. The reasons for the different behavior of the SA and SST models here and in Cecora et al. are not known. Figure 6 shows surface pressure coefficient for five turbulence models (SSG/LRR-RSM-w2012-SD results were essentially the same as SSG/LRR-RSM-w2012, therefore are not shown). All models except for WilcoxRSM-w2006 produced results in good agreement with the experiment at the outboard stations. At the two inboard stations, all the results were essentially the same, and showed some disagreement with experiment, possibly because of the boundary condition issues mentioned above. These inboard-station results are typical of other CFD results that use a symmetry boundary condition at the wing root.

Grid density effect is shown in Fig. 7 for the RSMs. For stations at and inboard of $2y/B = 0.90$, the influence of grid refinement was primarily a slight sharpening of the shock. Nearer the wing tip, at $2y/B = 0.96$ and 0.99, a larger influence was seen regarding the position of the shock as well as the pressure levels on the upper surface near the aft end of the wing. For SSG/LRR-RSM-w2012 (right column in the figure), results remained reasonable for all grid levels. However, for WilcoxRSM-w2006 (left column in the figure), grid refinement resulted in increased forward movement of the shock at the outboard stations. This suggests a possible reason for the disagreement between current SA and SST results and those of Cecora et al. Even though the current study used grids refined as high as 24 million cells (see picture of fine grid surface resolution near the wing tip in Fig. 8), the unstructured Cecora et al. grid with 4.6 million cells appeared to be very highly clustered near the tip. Therefore, it is possible that additional grid refinement in this region (beyond

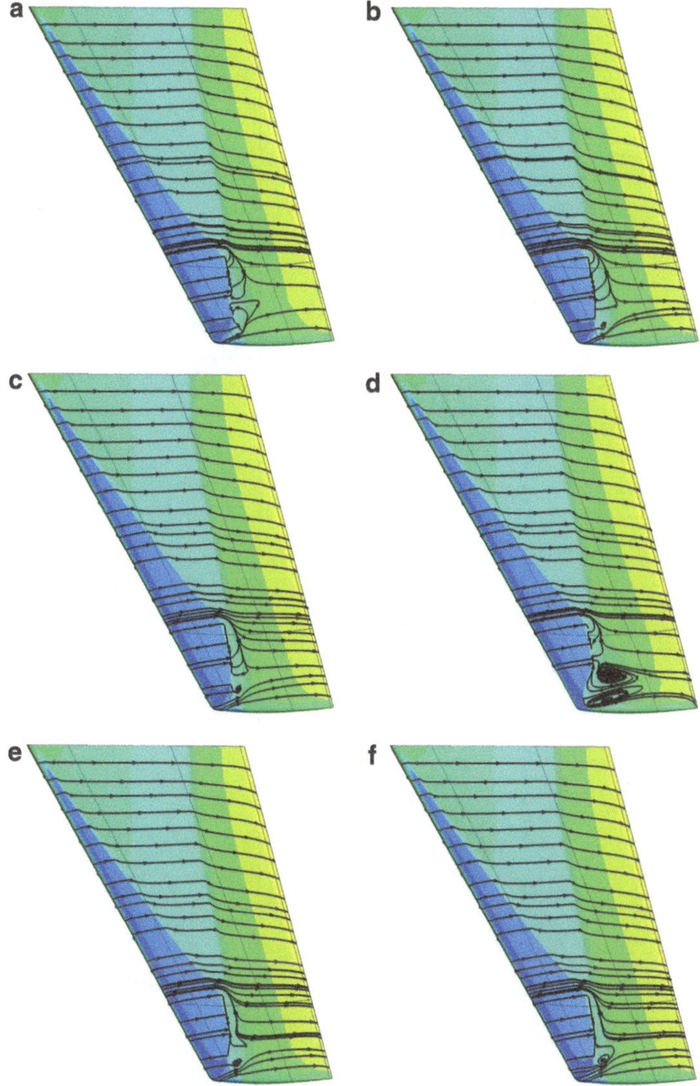

Fig. 5 Upper surface streamlines and pressure coefficient contours for the ONERA M6 wing at $\alpha = 4.08^\circ$ on $865 \times 193 \times 145$ grid (*flow is from left to right*). (**a**) SA model. (**b**) SST model. (**c**) EASMko2003-S model. (**d**) WilcoxRSM-w2006 model. (**e**) SSG/LRR-RSM-w2012 model. (**f**) SSG/LRR-RSM-w2012-SD model

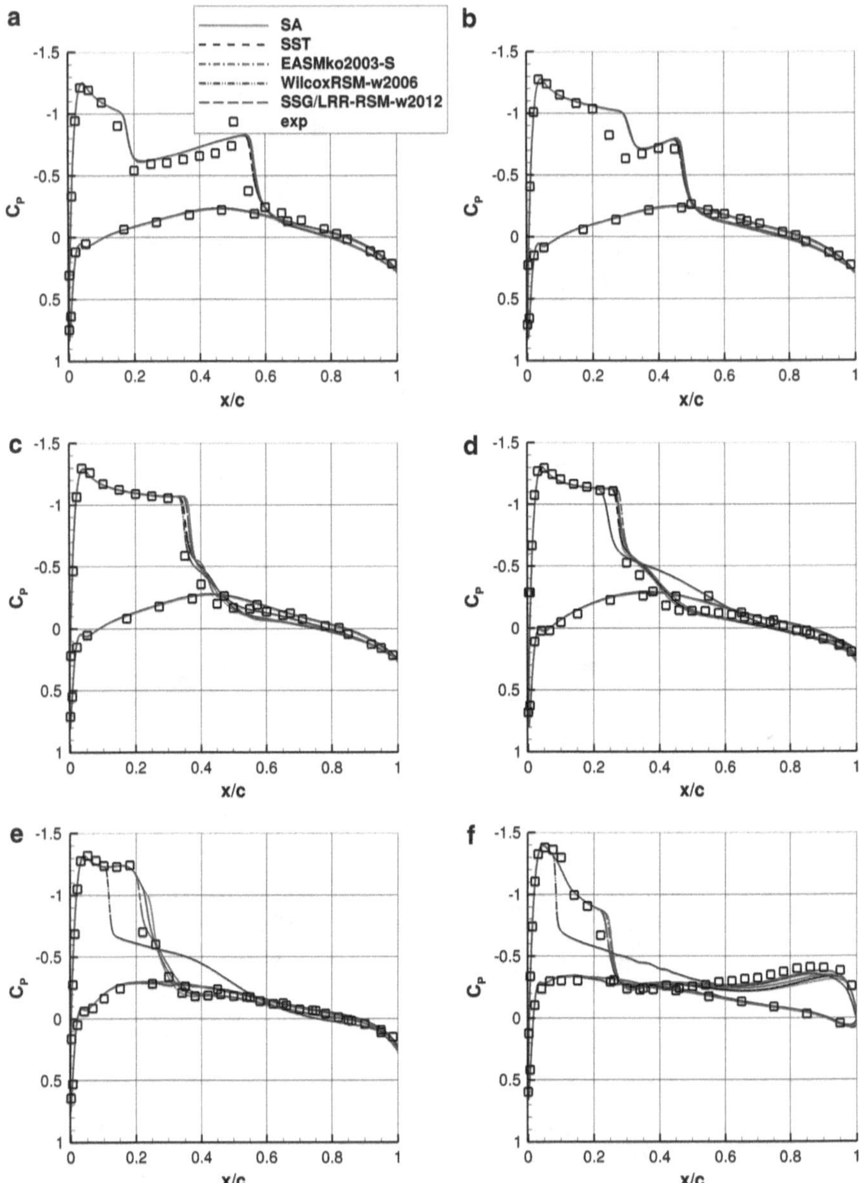

Fig. 6 Surface pressure coefficients for the ONERA M6 wing at $\alpha = 4.08°$ on $865 \times 193 \times 145$ grid. (**a**) Span location $2y/B = 0.44$. (**b**) Span location $2y/B = 0.65$. (**c**) Span location $2y/B = 0.80$. (**d**) Span location $2y/B = 0.90$. (**e**) Span location $2y/B = 0.96$. (**f**) Span location $2y/B = 0.99$

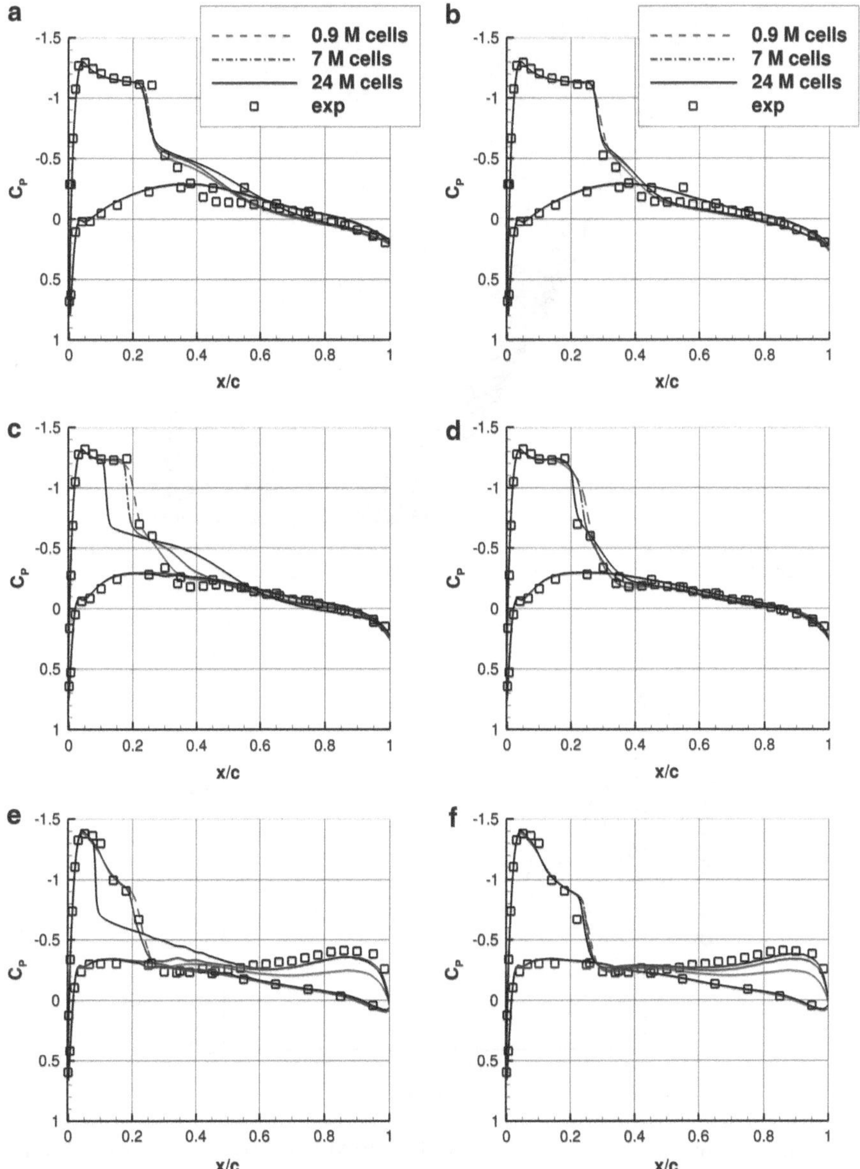

Fig. 7 Effect of grid on RSM surface pressure coefficients for the ONERA M6 wing at $\alpha = 4.08°$. (**a**) WilcoxRSM-w2006, $2y/B = 0.90$. (**b**) SSG/LRR-RSM-w2012, $2y/B = 0.90$. (**c**) WilcoxRSM-w2006, $2y/B = 0.96$. (**d**) SSG/LRR-RSM-w2012, $2y/B = 0.96$. (**e**) WilcoxRSM-w2006, $2y/B = 0.99$. (**f**) SSG/LRR-RSM-w2012, $2y/B = 0.99$

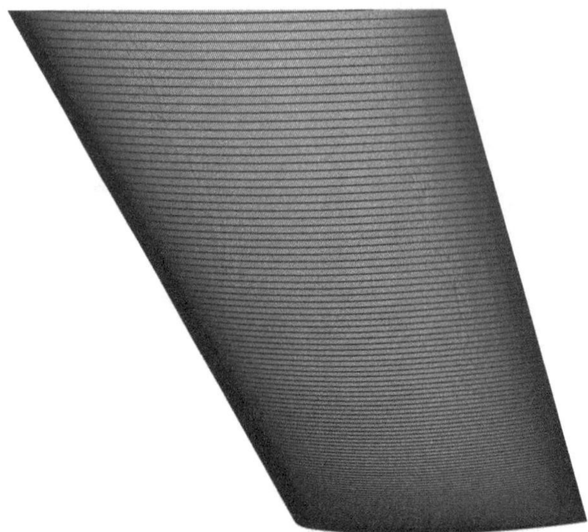

Fig. 8 Upper surface of 865 × 193 × 145 ONERA M6 grid near wing tip

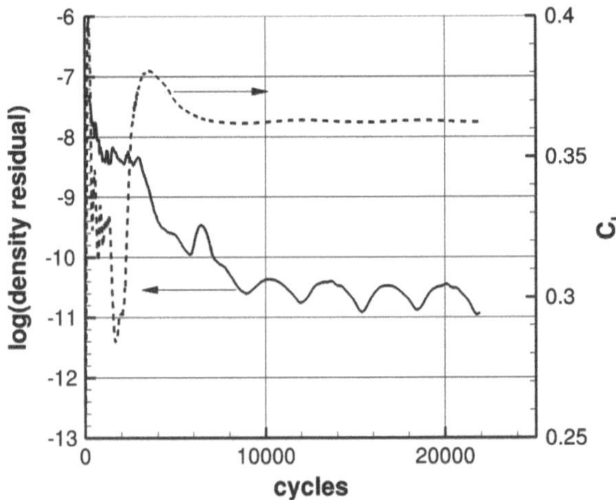

Fig. 9 Example of convergence behavior of SSG/LRR-RSM-w2012 model in CFL3D for ONERA M6 wing on 865 × 193 × 145 grid

the current fine grid) may be necessary to see a solution with massive separation when using SA or SST. In any case, the good agreement of current SSG/LRR-RSM-w2012 computations for this case with those in Cecora et al. using the same model are further confirmation of this particular model's ability to capture this flow.

Figure 9 shows a sample convergence behavior for the SSG/LRR-RSM-w2012 in CFL3D. On the finest ONERA M6 grid, the density residual was reduced about 3 orders of magnitude over approximately 12,000 multigrid cycles, then it leveled off and did not reduce much further over an additional 10,000 cycles. This allowed the lift to reach a quasi-steady state with lift oscillations between approximately 0.3624 and 0.3631 (less than 0.2 % change). Although not readily visible in this plot, it appeared that the lift oscillations would continue to decrease with additional multigrid cycles.

5.3 NASA Common Research Model

The NASA Common Research Model (CRM) has been used in the fourth and fifth Drag Prediction Workshops [12, 30]. The CRM is a wing-body configuration (the fourth workshop included a horizontal tail while the fifth did not). The CRM configuration was investigated in the current study because at angle-of-attack of $\alpha = 4°$, workshop participants reported a wide range of wing-root separation bubble sizes. Yamamoto et al. [32] was one of the first participants to demonstrate that the *form* of the turbulence model (linear vs. non-linear) had a significant effect on the prediction of this bubble size. They found that use of a quadratic constitutive relation in conjunction with SA reduced the size of the separation bubble significantly, in better agreement with experimental evidence. A description of SA-QCR2000 can be found in Spalart [26].

Here, computations were conducted on a structured grid provided by JAXA from the fourth workshop, with approximately 11 million grid cells. Flow conditions were $\alpha = 4°$, $M = 0.85$, $Re_{\text{MAC}} = 5$ million. For this case, the SSG/LRR-RSM-w2012 would not run successfully; it was necessary to make use of the SSG/LRR-RSM-w2012-SD version of the model (with simple diffusion) instead. Results are summarized in Fig. 10. As seen in Fig. 10a,b both SA and SST (which are linear Boussinesq models) produced fairly large wing-root separation bubbles. But all the models that make use of nonlinear constitutive relations—SA-QCR2000, EASMk02003-S, WilcoxRSM-w2006, and SSG/LRR-RSM-w2012-SD—yielded very small bubbles. This set of computations adds further evidence to the realization that it is imperative to include nonlinear behavior in RANS turbulence models when computing corner-type flows such as this.

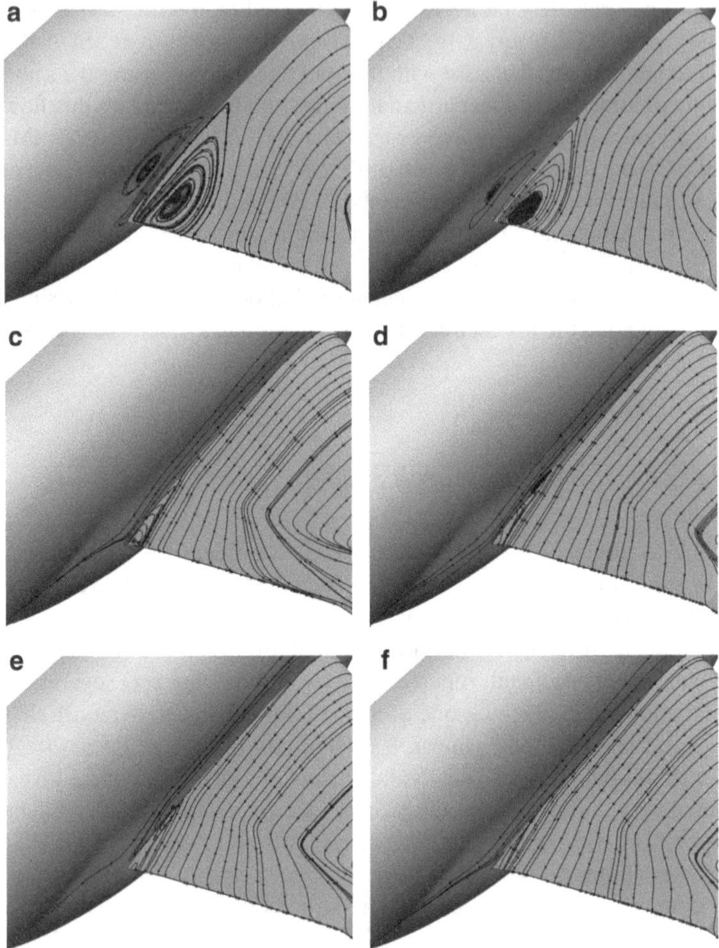

Fig. 10 Surface streamlines in wing-body juncture region of NASA CRM, $\alpha = 4°$ (*flow is from top right to bottom left*). (**a**) SA model. (**b**) SST model. (**c**) SA-QCR2000 model. (**d**) EASMko2003-S model. (**e**) WilcoxRSM-w2006 model. (**f**) SSG/LRR-RSM-w2012-SD model

6 Conclusions

Several ω-based second-moment Reynolds stress turbulence models that have been recently implemented in NASA CFD codes were tested on three different separated aerodynamic flows. For the 2-D flow over the NASA wall-mounted hump, the RSMs (like other RANS models) failed to properly account for the high turbulence activity in the separated shear layer. As a result, reattachment occurred too late. Comparing RSMs, the SSG/LRR-RSM-w2012 and SSG/LRR-RSM-w2012-SD models yielded better results (earlier reattachment) than WilcoxRSM-w2006. The former models were also somewhat better than SA or SST in their reattachment positions, but the

RSMs suffered from a well-known problem of unrealistic streamline back bending near reattachment. This problem, if fixed, would likely influence the reattachment location.

For computations of shock-induced separated flow over the ONERA M6 wing, the SSG/LRR-RSM-w2012 and SSG/LRR-RSM-w2012-SD models both yielded good results compared with experiment. The WilcoxRSM-w2006 model, however, predicted the shock too far forward with a larger region of separation near the wing tip as the grid was refined. In the current computations, the SA and SST models both agreed well with the SSG/LRR-RSM-w2012 computations and with experiment. This result is at odds with an earlier report by Cecora et al., who showed SA and SST to incorrectly produce massive separation at these conditions. To try to resolve this discrepancy, further grid refinement beyond what was done here may be warranted.

Finally, computations on the NASA CRM configuration substantiated the notion that inclusion of nonlinear effects is important when computing corner flows like this. SA and SST, linear models that make use of the Boussinesq assumption, were shown to produce large wing-root corner separation bubbles, contrary to experimental evidence. The RSMs, which obviously account for normal stress differences, yielded small wing-root corner separation bubbles. However, nonlinear effects can also be easily accounted for in one- and two-equation models, such as in SA-QCR2000 and EASMko2003-S. These models also produced small bubbles for this case.

In conclusion, the SSG/LRR-RSM-w2012 model was demonstrated to be generally better than WilcoxRSM-w2006 for the separated flows tested. However, definitive benefits of the RSM over simpler one- and two-equation models have not been established yet. Although the RSM automatically includes important nonlinear behaviors, these can be easily added (although perhaps less rigorously) in one- and two-equation models. And, like other RANS models, the current RSMs still cannot account for the high levels of turbulence inherent in some separated shear layers. Nonetheless, the RSMs may provide a convenient testing bed for possible future turbulence model improvements aimed toward better predictions of separated flows.

Acknowledgements The author acknowledges Dr. John Vassberg of the Boeing Company for providing a set of refined grids for the ONERA M6 wing. Dr. Bernhard Eisfeld of DLR is recognized for his generous time spent collaborating with the author in verifying and testing the SSG/LRR-RSM-w2012 model. The author also acknowledges Dr. Xudong Xiao of Corvid Technologies, who subcontracted under Dr. Hassan on NASA Cooperative Agreement NNX11AI56A, for his expertise in implementation of the Wilcox RSM into two NASA codes.

References

1. Anderson WK, Bonhaus DL (1994) An implicit upwind algorithm for computing turbulent flows on unstructured grids. Comput Fluids 23(1):1–22
2. Cecora R-D, Eisfeld B, Probst A, Crippa S, Radespiel R (2012) Differential Reynolds stress modeling for aeronautics. AIAA paper 2012-0465, 50th AIAA aerospace sciences meeting, Nashville, 9–12 Jan 2012

3. Daly BJ, Harlow FH (1970) Transport equations in turbulence. Phys Fluids 13(11):2634–2649
4. Eisfeld B, Brodersen O (2005) Advanced turbulence modelling and stress analysis for the DLR-F6 condifguration. AIAA paper 2005-4727, 23rd AIAA applied aerodynamics conference, Toronto, 6–9 June 2005
5. FUN3D Development Team (2014) FUN3D. http://fun3d.larc.nasa.gov. Accessed 20 Aug 2014
6. German Aerospace Center (2014) Homepage of DLR TAU Code. http://tau.dlr.de. Accessed 20 Aug 2014
7. Hanjalić K, Jakirlić S (1998) Contribution towards the second-moment closure modelling of separating turbulent flows. Comput Fluids 27(2):137–156
8. Hanjalić K, Launder BE (1972) A Reynolds stress model of turbulence and its application to thin shear flows. J Fluid Mech 52(4):609–638
9. Jakirlić S, Hanjalić K (2002) A new approach to modelling near-wall turbulence energy and stress dissipation. J Fluid Mech 539:139–166
10. Krist SL, Biedron RT, Rumsey CL (1998) CFL3D user's manual (version 5.0). NASA/TM-1998-208444 June
11. Launder BE, Reece GJ, Rodi W (1975) Progress in the development of a Reynolds-stress turbulence closure. J Fluid Mech 68(3):537–566
12. Levy DW, Laflin KR, Tinoco EN, Vassberg JC, Mani M, Rider B, Rumsey CL, Wahls RA, Morrison JH, Brodersen OP, Crippa S, Mavriplis DJ, Murayama M (2014) Summary of data from the fifth computational fluid dynamics drag prediction workshop. J Aircr 51(4):1194–1213
13. Menter FR (1994) Two-equation eddy-viscosity turbulence models for engineering applications. AIAA J 32(8):1598–1605
14. Oberkampf WL, Trucano TG (2008) Verification and validation benchmarks. Nucl Eng Des 238:716–743
15. Rodio JJ, Xiao X, Hassan HA, Rumsey CL (2014) NASA trapezoidal wing simulation using stress-ω and one- and two-equation turbulence models. AIAA paper 2014-0404, AIAA SciTech, National Harbor, 13–17 Jan 2014
16. Roe PL (1981) Approximate Riemann solvers, parameter vectors, and difference schemes. J Comput Phys 43:357–372
17. Rumsey CL (2009) Successes and challenges for flow control simulations. Int J Flow Control 1(1):1–27
18. Rumsey CL (2014) CFD Validation of synthetic jets and turbulent separation control. http://cfdval2004.larc.nasa.gov. Accessed 20 Aug 2014
19. Rumsey CL (2014) CFL3D Version 6. http://cfl3d.larc.nasa.gov. Accessed 20 Aug 2014
20. Rumsey CL (2014) Turbulence modeling resource. http://turbmodels.larc.nasa.gov. Accessed 20 Aug 2014
21. Rumsey CL, Gatski TB (2003) Summary of EASM turbulence models in CFL3D with validation test cases. NASA/TM-2003-212431
22. Rumsey CL, Gatski TB, Sellers WL III, Vatsa VN, Viken SA (2006) Summary of the 2004 computational fluid dynamics validation workshop on synthetic jets. AIAA J 44(2):194–207
23. Schmitt V, Sharpin F (1979) Pressure distributions on the ONERA-M6-wing at transonic mach numbers. In: Experimental data base for computer program assessment. AGARD-AR-138, B1
24. Schwamborn D, Gardner A, von Geyr H, Krumbein A, Ludeke A, Sturmer A (2008) Development of the TAU-code for aerospace applications. In: 50th NAL international conference on aerospace science and technology, 2008-06-26–2008-06-28, Bangalore
25. Shir CC (1973) A preliminary numerical study of atmospheric turbulent flows in the idealized planetary boundary layer. J Atmos Sci 30(10):1327–1339
26. Spalart PR (2000) Strategies for turbulence modelling and simulation. Int J Heat Fluid Flow 21:252–263
27. Spalart PR, Allmaras SR (1994) A one-equation turbulence model for aerodynamic flows. Rech Aerosp 1:5–21
28. Speziale CG, Sarkar S, Gatski TB (1991) Modelling the pressure-strain correlation of turbulence: an invariant dynamical systems approach. J Fluid Mech 227:245–272

29. Thompson KB, Hassan HA (2014) Simulation of a variety of wings using a Reynolds stress model. AIAA paper 2014-2192, AIAA Aviation, Atlanta, 16–20 Jun 2014
30. Vassberg JC, Tinoco EN, Mani M, Rider B, Zickuhr T, Levy DW, Brodersen OP, Eisfeld B, Crippa S, Wahls RA, Morrison JH, Mavriplis DJ, Murayama M (2014) Summary of the fourth AIAA computational fluid dynamics drag prediction workshop. J Aircr 51(4):1070–1089
31. Wilcox DC (2006) Turbulence modeling for CFD, 3rd edn. DCW Industries, La Cañada
32. Yamamoto K, Tanaka K, Murayama M (2010) Comparison study of drag prediction for the 4th CFD drag prediction workshop using structured and unstructured mesh methods. AIAA paper 2010–4222, 28th AIAA Applied aerodynamics conference, Chicago, 28 June–1 July 2010

Separated Flow Prediction Around a 6:1 Prolate Spheroid Using Reynolds Stress Models

Yair Mor-Yossef

Abstract A numerical study of the separated flow about a 6:1 prolate spheroid at high-angle of attack using state-of-the-art Reynolds stress models is presented. The convective fluxes of the mean-flow and the Reynolds stress model equations are approximated by a third-order upwind biased MUSCL scheme. The diffusive flux is approximated by second-order central differencing based on a full-viscous stencil. The objective is to evaluate the applicability of RSM to realistic high-Reynolds separated flows. Comprehensive comparisons of the boundary layer velocity profile and of the Reynolds stress tensor components against the experimental data are presented. A very good agreement between the experimental measurements and calculated boundary layer velocity profiles is obtained. However, only reasonable agreement is obtained for the Reynolds stress components. It is shown that the common first-order upwind approximation of the Reynolds stress model convective flux alone may adversely affect the accuracy of the solution.

1 Introduction

Flows around bluff and slender bodies comprise a variety of coexisting complex flow phenomena, such as flow separation, cross-flow separation, highly skewed boundary layer, and transitional flow. A representative case is the flow around prolate spheroid. A prolate spheroid is a three-dimensional body with two length scales, one plane of symmetry, and one axis of symmetry. The ratio between the semi-major and semi-minor axes, i.e., the aspect ratio, is a measure of departure from a spherical body. Spheroids with aspect ratios 8:1, 6:1, and 3:1 can be considered as simplified models of submarines, unmanned underwater vehicles, missiles, airships, etc. When the aspect ratio is relatively high and the major axis is aligned with the flows the prolate spheroid behaves as a slender body.

Despite its simple geometry, the flow around a prolate spheroid at incidence exhibits complex flow features, such as cross-flow separation, streamline curvature,

Y. Mor-Yossef (✉)
ISCFDC, Caesarea Industrial Park, Caesarea, Israel
e-mail: yairm@iscfdc.co.il

B. Eisfeld (ed.), *Differential Reynolds Stress Modeling for Separating Flows in Industrial Aerodynamics*, Springer Tracts in Mechanical Engineering,
DOI 10.1007/978-3-319-15639-2_3

Fig. 1 Cross-flow separation and stream-wise vortices presented by streamlines colored by pressure on a prolate spheroid at incidence of $\alpha = 20^\circ$, $Re_\infty = 4.2 \times 10^6$, $M_\infty = 0.2$ (the streamlines are mirrored for visualization purposes only)

the formation and evolution of free-vortex sheets, and stream-wise vortices. Some of these are reflected in Fig. 1. For moderate incidence angles, the flow separates from the leeward side of the prolate spheroid and rolls up into coherent longitudinal symmetric vortices. A pair of primary vortices that is usually accompanied by at least one pair of secondary vortices.

For two-dimensional flows, analysis of flow separation provides a detailed description of the conditions influencing many separated flows. In contrast, three-dimensional separated flow analysis presents a great challenge. In three-dimensional flows, separation characteristics can be sensitive to the body geometry and angle of attack and Reynolds number, among other factors. Flow reversal and vanishing of the shear stress are two well-known effects that may not accompany three-dimensional separations. Another flow feature that characterizes the flow around a prolate spheroid is transition. Generally, up to a certain Reynolds number, and at a low Mach-number, only transition due to Tollmien–Schlichting (TS) instabilities occur on the surface of the prolate spheroid. At a higher Reynolds number and at a moderate incidence, transition is caused by both TS and crossflow instabilities. At still higher incidence, Reynolds number, and Mach number, transition is dominated by crossflow instabilities. Bearing in mind the above-mentioned flow phenomena, it is clear that accurate modeling of the flow about a prolate spheroid is extremely challenging.

A series of experimental studies about a 6:1 prolate spheroid have been conducted by the group at Virginia Polytechnic Institute (VPI) [6, 14, 27]. Their comprehensive experimental data reveal salient physics of the flow by offering an invaluable dataset. Past numerical studies about the 6:1 prolate spheroid by the

VPI group include the use of Reynolds-averaged Navier–Stokes (RANS) turbulence models [7, 18, 25, 26], hybrid RANS/LES models [7, 29], and Large Eddy Simulations (LES) [28]. While the LES approach and even the hybrid RANS/LES approach are prohibitively expensive for practical engineering applications, RANS turbulence models are a compromise between accuracy and affordability.

Among the RANS turbulence models, Reynolds-stress models (RSM) are perceived as the most advanced ones. A clear advantage of RSM (over first-order closure models) is that the production term does not require approximations. It is the production term that is primarily responsible for the anisotropy and the selective response of turbulence to different strain types. Hence, it is expected that this higher level of modeling, representing more elaborate physics, would be beneficial in terms of accurate flow predictions. Among the past studies which employ RANS turbulence models for the numerical simulations about the VPI prolate spheroid, only Kim et al. [18] use Reynolds stress models. However, they use a relatively coarse grid and employ a wall function.

In recent years, there are renewed efforts in the development [9, 13, 16, 19] and interest in the application [3–5, 12, 17, 21, 22] of Reynolds stress models, for practical engineering applications. As a part of this effort, the aim of the present work is to simulate the flow about the 6:1 prolate spheroid in accordance with the VPI experimental flow conditions, using Reynolds stress models.

2 Governing Equations

The governing equations are obtained by Favre-averaging the Navier–Stokes equations (RANS) and modeling the Reynolds stress. The unknown Favre-averaging Reynolds stress tensor is modeled in this work via a second-moments closure model. In the proceeding equations, the symbol ($\bar{\ }$) indicates non-weighted averaging, the symbol ($\tilde{\ }$) signifies Favre averaging, and the symbol ($''$) denotes Favre fluctuations.

2.1 Mean-Flow Equations

The Favre averaged mean-flow equations may be expressed in Cartesian coordinates as follows:

$$\frac{\partial \bar{\rho}}{\partial t} + \frac{\partial \left(\bar{\rho}\tilde{u}_k\right)}{\partial x_k} = 0 \tag{1}$$

$$\frac{\partial \bar{\rho}\tilde{u}_i}{\partial t} + \frac{\partial \left(\bar{\rho}\tilde{u}_i\tilde{u}_k\right)}{\partial x_k} = -\frac{\partial \bar{p}}{\partial x_i} + \frac{\partial(\bar{\tau}_{ik} - \bar{\rho}\tilde{\mathfrak{R}}_{ik})}{\partial x_k} \tag{2}$$

$$\frac{\partial \tilde{E}}{\partial t} + \frac{\partial[(\tilde{E} + \bar{p})\tilde{u}_k]}{\partial x_k} = \frac{\partial[(\bar{\tau}_{ik} - \bar{\rho}\tilde{\mathfrak{R}}_{ik})\tilde{u}_i]}{\partial x_k} - \frac{\partial[\bar{q}_k + (\bar{q}_t)_k]}{\partial x_k} \tag{3}$$

where t denotes the time, $x_i = [x, y, z]$ are the Cartesian coordinates, and $\tilde{u}_i = [\tilde{u}, \tilde{v}, \tilde{w}]$ are the Cartesian velocity vector components. The fluid density is denoted by $\bar{\rho}$, $\bar{p}$ denotes the pressure, and the total energy is denoted by $\tilde{E}$. The viscous stress tensor, $\bar{\tau}_{ij}$, is given as

$$\bar{\tau}_{ij} = \bar{\mu}\left(\frac{\partial \tilde{u}_i}{\partial x_j} + \frac{\partial \tilde{u}_j}{\partial x_i} - \frac{2}{3}\frac{\partial \tilde{u}_k}{\partial x_k}\delta_{ij}\right) \tag{4}$$

The tensor $\tilde{\mathfrak{R}}_{ij} = \widetilde{u_i'' u_j''}$ is the Reynolds-stress tensor. $\bar{q}_i$ and $(\bar{q}_t)_i$ are the molecular and turbulent heat fluxes, respectively, modeled using Fourier's law:

$$\bar{q}_i = -\bar{\kappa}\bar{T}_{x_i} \tag{5}$$

$$(\bar{q}_t)_i = -\bar{\kappa}_t\bar{T}_{x_i} \tag{6}$$

with $\bar{T}$ denoting the temperature and $\bar{\kappa} = c_p\bar{\mu}/Pr$ and $\bar{\kappa}_t = c_p\mu_t/Pr_t$ are the molecular and turbulent heat conductivities, respectively. The term $\bar{\mu}$ denotes the molecular viscosity, calculated using Sutherland's law, and μ_t is the turbulent viscosity based on a linear eddy-viscosity model. The term c_p is the specific heat capacity at constant pressure and $Pr = 0.72$, $Pr_t = 0.9$ are the molecular and turbulent Prandtl numbers, respectively. The mean-flow equations are closed using the equation of state for a perfect gas, given by:

$$\bar{p} = (\gamma - 1)\left[\tilde{E} - \frac{1}{2}\bar{\rho}\left(\tilde{u}^2 + \tilde{v}^2 + \tilde{w}^2\right)\right] \tag{7}$$

where γ is the ratio of specific heats $\left(c_p/c_v\right)$, set to $\gamma = 1.4$. Note that the contribution of the turbulent diffusion to the total energy transport equation is neglected, as well as the contribution of the turbulent kinetic energy to the total energy.

2.2 Reynolds-Stress Model Equations

Reynolds-stress models use the exact equations for the transport of Reynolds stresses obtained by taking velocity-weighted moments of the Navier–Stokes equations and neglecting density fluctuations. The general form of a Reynolds stress model is given by

$$\frac{\partial(\bar{\rho}\tilde{\mathfrak{R}}_{ij})}{\partial t} + \frac{\partial(\bar{\rho}\tilde{\mathfrak{R}}_{ij}\tilde{u}_k)}{\partial x_k} = P_{ij} + \Pi_{ij} - \varepsilon_{ij} + D^{\nu}_{ij} + D^{t}_{ij} + D^{P}_{ij} \tag{8}$$

where the production term

$$P_{ij} = -\bar{\rho}\tilde{\mathfrak{R}}_{ik}\frac{\partial \tilde{u}_j}{\partial x_k} - \bar{\rho}\tilde{\mathfrak{R}}_{jk}\frac{\partial \tilde{u}_i}{\partial x_k} \tag{9}$$

is exact. The term D^{ν}_{ij} denotes the molecular diffusion given as

$$D^{\nu}_{ij} = \frac{\partial}{\partial x_k}\left(\bar{\mu}\frac{\partial \tilde{\mathfrak{R}}_{ij}}{\partial x_k}\right) \tag{10}$$

The remaining terms on the right-hand side of Eq. (8) require modeling; Π_{ij} is the pressure-strain correlation; ε_{ij} is the turbulent dissipation; D^{t}_{ij} is the turbulent diffusion; and D^{P}_{ij} is the pressure diffusion.

For the reason of a consistent notation, the following definition applies: The Reynolds-stress anisotropy tensor is given by

$$a_{ij} = \frac{\tilde{\mathfrak{R}}_{ij}}{k} - \frac{2}{3}\delta_{ij} \tag{11}$$

where $k = 0.5\tilde{\mathfrak{R}}_{jj}$. Here, Lumley's stress-flatness parameter, which varies between unity in isotropic turbulence regions and zero as the turbulence approaches a two-component limit, is denoted by A and defined as:

$$A = 1 - \frac{9}{8}(A_2 - A_3), \quad A_2 = a_{ij}a_{ij}, \quad A_3 = a_{ij}a_{jk}a_{ki} \tag{12}$$

The mean-strain tensor, S_{ij}, and the averaged rotation tensor, Ω_{ij}, are defined as:

$$S_{ij} = \frac{1}{2}\left(\frac{\partial \tilde{u}_i}{\partial x_j} + \frac{\partial \tilde{u}_j}{\partial x_i}\right), \quad \Omega_{ij} = \frac{1}{2}\left(\frac{\partial \tilde{u}_i}{\partial x_j} - \frac{\partial \tilde{u}_j}{\partial x_i}\right) \tag{13}$$

The turbulence Reynolds number, Re_t, and the turbulence length scale, l_t, are given as

$$Re_t = \frac{k^2}{\bar{\nu}\varepsilon}, \quad l_t = \frac{k^{3/2}}{\varepsilon} \tag{14}$$

and in a similar manner, the turbulence Reynolds number, Re^*_t, and turbulence length scale, l^*_t, which are based on the homogeneous dissipation rate of k, and ε^*, are given as

$$Re^*_t = \frac{k^2}{\bar{\nu}\varepsilon^*}, \quad l^*_t = \frac{k^{3/2}}{\varepsilon^*} \tag{15}$$

with

$$\varepsilon^* = \varepsilon - 2\bar{\nu}\left(\frac{\partial k^{1/2}}{\partial x_j}\right)^2 \tag{16}$$

Three Reynolds stress models were tested in the current work: The MCL model, proposed by Batten et al. [2], which is a modified version of the Craft-Launder closure [8], designed for compressible aerodynamics applications; The hybrid SSG/LRR-ω model developed by Eisfeld [9]; The GLVY model, recently developed by Gerolymos et al. [13].

2.2.1 MCL Model

The MCL model employs a cubic pressure-strain model developed by Fu [10] in conjunction with coefficients and inhomogeneity corrections modified from those proposed by Craft and Launder [8]

$$\Pi_{ij} = \Pi_{ij,1} + \Pi_{ij,2} + \Pi_{ij,1}^{inh} + \Pi_{ij,2}^{inh} \tag{17}$$

$$\Pi_{ij,1} = -C_1\bar{\rho}\varepsilon^*\left[a_{ij} + C_1'\left(a_{ik}a_{jk} - \frac{1}{3}A_2\delta_{ij}\right)\right] - \bar{\rho}\varepsilon^* A^{1/2}a_{ij} \tag{18}$$

$$\begin{aligned}
\Pi_{ij,2} = {} & -0.6\left(P_{ij} - \frac{1}{3}P_{kk}\delta_{ij}\right) + 0.3a_{ij}P_{kk} - 0.2\frac{\bar{\rho}}{k}\left[\tilde{\mathfrak{R}}_{kj}\tilde{\mathfrak{R}}_{li}\left(\frac{\partial\tilde{u}_k}{\partial x_l} + \frac{\partial\tilde{u}_l}{\partial x_k}\right)\right. \\
& \left. - \mathfrak{R}_{lk}\left(\tilde{\mathfrak{R}}_{ik}\frac{\partial\tilde{u}_j}{\partial x_l} + \tilde{\mathfrak{R}}_{jk}\frac{\partial\tilde{u}_i}{\partial x_l}\right)\right] - C_2\left[A_2\left(P_{ij} - D_{ij}\right) + 3a_{mi}a_{nj}\left(P_{mn} - D_{mn}\right)\right] \\
& + C_2'\left\{\left(\frac{7}{15} - \frac{1}{4}A_2\right)\left(P_{ij} - \frac{1}{3}P_{kk}\delta_{ij}\right)\right] + 0.1\left[a_{ij} - \frac{1}{2}\left(a_{ik}a_{kj}\right.\right. \\
& \left.\left. - \frac{1}{3}A_2\delta_{ij}\right)\right]P_{kk} - 0.05a_{ij}a_{lk}P_{kl} + 0.1\frac{1}{k}\left(\tilde{\mathfrak{R}}_{im}P_{mj} + \tilde{\mathfrak{R}}_{jm}P_{mi}\right. \\
& \left. - \frac{2}{3}\tilde{\mathfrak{R}}_{lm}P_{ml}\delta_{ij}\right) + 0.1\frac{1}{k^2}\left(\tilde{\mathfrak{R}}_{li}\tilde{\mathfrak{R}}_{kj} - \frac{1}{3}\tilde{\mathfrak{R}}_{lm}\tilde{\mathfrak{R}}_{km}\delta_{ij}\right)\left[6D_{lk}\right. \\
& \left. + 13\bar{\rho}k\left(\frac{\partial\tilde{u}_l}{\partial x_k} + \frac{\partial\tilde{u}_k}{\partial x_l}\right)\right] + 0.2\frac{1}{k^2}\mathfrak{R}_{li}\tilde{\mathfrak{R}}_{kj}\left(D_{lk} - P_{lk}\right)\Big\}
\end{aligned} \tag{19}$$

$$\Pi^{inh}_{ij,1} = \frac{\bar{\rho}\varepsilon}{k} f_{\omega 1} \left(\tilde{\Re}_{lk} d^A_l d^A_k \delta_{ij} - \frac{3}{2} \tilde{\Re}_{ik} d^A_j d^A_k - \frac{3}{2} \tilde{\Re}_{jk} d^A_i d^A_k \right)$$
$$+ \frac{\bar{\rho}\varepsilon}{k^2} f_{\omega 2} \left(\tilde{\Re}_{mn} \tilde{\Re}_{ml} d^A_n d^A_l \delta_{ij} - \frac{3}{2} \tilde{\Re}_{im} \tilde{\Re}_{ml} d^A_j d^A_l \right.$$
$$\left. - \frac{3}{2} \tilde{\Re}_{jm} \tilde{\Re}_{ml} d^A_i d^A_l \right) \tag{20}$$

$$\Pi^{inh}_{ij,2} = f_I \bar{\rho} k \frac{\partial \tilde{u}_l}{\partial x_n} d_l d_n \left(d_i d_j - \frac{1}{3} d_k d_k \delta_{ij} \right) \tag{21}$$

with

$$D_{ij} = -\bar{\rho} \tilde{\Re}_{ik} \frac{\partial \tilde{u}_k}{\partial x_j} - \bar{\rho} \tilde{\Re}_{jk} \frac{\partial \tilde{u}_k}{\partial x_i} \tag{22}$$

In the preceding equations, the gradient-indicators vectors are

$$d_i = \frac{N_i}{0.5 + \sqrt{N_k N_k}}, \qquad d^A_i = \frac{N^A_i}{0.5 + \sqrt{N^A_k N^A_k}} \tag{23}$$

with the gradients and the turbulence length scale given by

$$N_i = \frac{\partial (l_t A)}{\partial x_i}, \qquad N^A_i = \frac{\partial (l_t \sqrt{A})}{\partial x_i} \tag{24}$$

The pressure-strain correlation model coefficients are:

$$\begin{aligned}
C_1 &= 3.2 f_A A_2^{1/2} \min\left[(Re_t/160)^2, 1 \right], \quad C_1' = 1.1, \\
C_2 &= \min\left\{ 0.55\left[1 - \exp\left(-A^{3/2} Re_t/100 \right) \right], 3.2A/(1+S) \right\}, \\
C_2' &= \min\left(0.6, A^{1/2} \right) + 3.5\,(S - \Omega)/(3 + S + \Omega) - 4 \min\left(S_I, 0 \right), \\
f_{\omega 1} &= 2.35\left(1 - A^{1/2} \right) \min\left\{ 1, \max\left[1 - (Re_t - 55)/70, 0 \right] \right\}, \\
f_{\omega 2} &= 0.6 A_2 \left(1 - A^{1/2} \right) \min\left\{ 1, \max\left[1 - (Re_t - 50)/85, 0 \right] \right\} + 0.1, \\
f_I &= 3 f_A, \\
f_A &= \begin{cases} (A/14)^{1/2}, & A < 0.05 \\ A/7^{1/2}, & 0.05 \le A < 0.7 \\ A^{1/2}, & A \ge 0.7 \end{cases}
\end{aligned} \tag{25}$$

with the following definitions

$$S = \frac{k}{\varepsilon}\sqrt{2S_{ij}S_{ij}}, \quad \Omega = \frac{k}{\varepsilon}\sqrt{2\Omega_{ij}\Omega_{ij}}, \quad S_I = \sqrt{6}S_{ij}S_{jk}S_{ki}/\left(S_{ln}S_{ln}\right)^{3/2} \tag{26}$$

The dissipation tensor is formulated using a blend of isotropic, ε, and wall limiting values, ε'_{ij}, with additional term, ε''_{ij}, to account for the dip in the shear-stress dissipation rate in the buffer layer:

$$\varepsilon_{ij} = (1 - f_\varepsilon)\frac{\bar{\rho}\left(\varepsilon'_{ij} + \varepsilon''_{ij}\right)}{D} + \frac{2}{3}f_\varepsilon\bar{\rho}\varepsilon\delta_{ij} \tag{27}$$

with

$$\begin{aligned}\varepsilon'_{ij} &= \varepsilon\frac{\tilde{\mathfrak{R}}_{ij}}{k} + 2\bar{\nu}\frac{\tilde{\mathfrak{R}}_{ln}}{k}\frac{\partial\sqrt{k}}{\partial x_l}\frac{\partial\sqrt{k}}{\partial x_n}\delta_{ij} \\ &\quad + 2\bar{\nu}\left(\frac{\tilde{\mathfrak{R}}_{li}}{k}\frac{\partial\sqrt{k}}{\partial x_j}\frac{\partial\sqrt{k}}{\partial x_l} + \frac{\tilde{\mathfrak{R}}_{lj}}{k}\frac{\partial\sqrt{k}}{\partial x_i}\frac{\partial\sqrt{k}}{\partial x_l}\right)\end{aligned} \tag{28}$$

$$\varepsilon''_{ij} = f_R\varepsilon\left(2\frac{\tilde{\mathfrak{R}}_{lk}}{k}d_l^A d_k^A\delta_{ij} - \frac{\tilde{\mathfrak{R}}_{li}}{k}d_l^A d_j^A - \frac{\tilde{\mathfrak{R}}_{lj}}{k}d_l^A d_i^A\right) \tag{29}$$

and the following definitions and coefficients

$$D = \frac{\varepsilon'_{kk} + \varepsilon''_{kk}}{2\varepsilon}, \quad f_R = (1 - A)\min\left[\left(Re_t/180\right)^2, 1\right], \quad f_\varepsilon = \sqrt{A} \tag{30}$$

The turbulent diffusion model that is used is in accordance with the generalized gradient diffusion hypothesis

$$D_{ij}^t = \frac{\partial}{\partial x_k}\left(0.22\frac{k}{\varepsilon}\bar{\rho}\tilde{\mathfrak{R}}_{kl}\frac{\tilde{\mathfrak{R}}_{ij}}{\partial x_l}\right) \tag{31}$$

and the pressure diffusion is modeled as follows:

$$D_{ij}^P = -\frac{\tilde{\mathfrak{R}}_{ij}}{2k}\frac{\partial\widehat{pu}_k}{\partial x_k} \tag{32}$$

with

$$\begin{aligned}\widehat{pu}_k &= -\bar{\rho}\left(1 - A\right)\left(0.5d_k + 1.1d_k^A\right)\left(\bar{\nu}\varepsilon kAA_2\right)^{1/2}C_{pd}, \\ C_{pd} &= \left[1 + 2\exp\left(-Re_t/40\right)\right]A_2 + 0.4Re_t^{-1/4}\exp\left(-Re_t/40\right)\end{aligned} \tag{33}$$

The MCL model uses an equation of the homogeneous dissipation rate, ε^*,

$$\frac{\partial \bar{\rho}\varepsilon^*}{\partial t} + \frac{\partial \bar{\rho}\varepsilon^* \tilde{u}_k}{\partial x_k} = \frac{\partial}{\partial x_l}\left[\left(\bar{\mu}\delta_{kl} + C_\varepsilon \bar{\rho}\tilde{\mathfrak{R}}_{kl}\frac{k}{\varepsilon}\right)\frac{\partial \varepsilon^*}{\partial x_l}\right] + C_{\varepsilon_1}\frac{\varepsilon}{2k}P_{kk} - C_{\varepsilon_2}\frac{\bar{\rho}\varepsilon^*\varepsilon}{k}$$
$$-C_{\varepsilon_3}\frac{(\varepsilon-\varepsilon^*)\,\bar{\rho}\varepsilon^*}{k} + C_{\varepsilon_4}\bar{\nu}\frac{k}{\varepsilon}\bar{\rho}\tilde{\mathfrak{R}}_{ij}\frac{\partial^2 \tilde{u}_k}{\partial x_i \partial x_l}\frac{\partial^2 \tilde{u}_k}{\partial x_j \partial x_l} + C_{\varepsilon_5}\frac{\bar{\rho}\varepsilon^*\varepsilon}{k}Y_p \tag{34}$$

with

$$Y_P = \min\left\{\max\left[F\,(F+1)^2, 0\right] 20\right\},$$
$$F = \frac{1}{2.55}\left[\sqrt{\left(\frac{\partial l_t}{\partial x_j}\right)\left(\frac{\partial l_t}{\partial x_j}\right)} - \left(\frac{\partial l_e}{\partial Y}\right)\right],$$
$$\frac{\partial l_e}{\partial Y} = 2.55[1 - \exp(-B_\varepsilon Re_t) + B_\varepsilon Re_t \exp(-B_\varepsilon Re_t)] \tag{35}$$

and the following calibration constants

$$B_\varepsilon = 0.1069, \quad C_{\varepsilon_1} = 1.44, \quad C_{\varepsilon_2} = 1.92, \quad C_{\varepsilon_3} = 1,$$
$$C_{\varepsilon_4} = 0.4, \quad C_{\varepsilon_5} = 0.2, \quad C_\varepsilon = 0.18\,. \tag{36}$$

2.2.2 SSG/LRR-ω Model

The idea behind the SSG/LRR-ω model is to blend the pressure-strain model of the LRR model, also used in the Wilcox Stress-ω model, and the SSG model. The rationale is to activate the LRR model in the near wall region while the SSG is activated further away. The blending is conducted based on the Menter [20] blending function

$$\Pi_{ij} = -\left(C_1\bar{\rho}\varepsilon + \frac{1}{2}C_1^* P_{kk}\right)a_{ij} + C_2\bar{\rho}\varepsilon\left(a_{ik}a_{kj} - \frac{1}{3}a_{kl}a_{kl}\delta_{ij}\right)$$
$$+\left(C_3 - C_3^*\sqrt{a_{kl}a_{kl}}\right)\bar{\rho}kS_{ij}^* + C_4\bar{\rho}k\left(a_{ik}S_{jk} + a_{jk}S_{ik} - \frac{2}{3}a_{kl}S_{kl}\delta_{ij}\right)$$
$$+C_5\bar{\rho}k\left(a_{ik}\Omega_{jk} + a_{jk}\Omega_{ik}\right) \tag{37}$$

where $S_{ij}^* = S_{ij} - \frac{1}{3}S_{kk}\delta_{ij}$.

Table 1 Bounding values of the SSG/LRR-ω model coefficients

	C_1	C_1^*	C_2	C_3	C_3^*	C_4	C_5	c_R
ϕ_R^{SSG}	3.4	1.8	4.2	0.8	1.3	1.25	0.4	$\frac{0.22}{C_\mu}$
ϕ_R^{LRR}	3.6	0	0	0.8	0	$\frac{18c_2^{\text{LRR}}+12}{11}$	$\frac{-14c_2^{\text{LRR}}+20}{11}$	0.75
$c_2^{\text{LRR}} = 0.52$								

The dissipation is modeled by an isotropic tensor:

$$\varepsilon_{ij} = \frac{2}{3}\bar{\rho}\varepsilon\delta_{ij} \tag{38}$$

where $\varepsilon = C_\mu k\omega$ with $C_\mu = 0.09$. The turbulent diffusion model that is used is in accordance with the generalized gradient diffusion hypothesis

$$D_{ij}^t = \frac{\partial}{\partial x_k}\left(c_R\frac{\bar{\rho}k}{\varepsilon}\tilde{\mathfrak{R}}_{kl}\frac{\partial\tilde{\mathfrak{R}}_{ij}}{\partial x_l}\right) \tag{39}$$

The coefficients $\phi_R = \left(C_1^*, C_3^*, C_1 - C_5, c_R\right)$ of the SSG/LRR-ω model are blended according to

$$\phi_R = F_1\phi_R^{\text{SSG}} + (1 - F_1)\,\phi_R^{\text{LRR}} \tag{40}$$

where the coefficients of ϕ_R^{SSG} and ϕ_R^{LRR} are given in Table 1.

The F_1 function, which was proposed by Menter [20], is given as:

$$F_1 = \tanh\left(\mathfrak{z}^4\right) \tag{41}$$

with the argument $\mathfrak{z}$ is given by:

$$\mathfrak{z} = \min\left[\max\left(\frac{\sqrt{k}}{C_\mu\omega d}, \frac{500\bar{\mu}}{\bar{\rho}\omega d^2}\right), \frac{4\sigma_\omega^{(\varepsilon)}\bar{\rho}k}{\sigma_d^{(\varepsilon)}\frac{\bar{\rho}}{\omega}\max\left(\frac{\partial k}{\partial x_k}\frac{\partial\omega}{\partial x_k}, 0\right)d^2}\right] \tag{42}$$

where d is the distance from the wall.

The SSG/LRR-ω model employs the equation of the specific turbulent dissipation rate, ω, according to the BSL model of Menter [20]

$$\frac{\partial\bar{\rho}\omega}{\partial t} + \frac{\partial\bar{\rho}\omega\tilde{u}_k}{\partial x_k} = \frac{\partial}{\partial x_k}\left[\left(\bar{\mu} + \sigma_\omega\frac{\bar{\rho}k}{\omega}\right)\frac{\partial\omega}{\partial x_k}\right] + \alpha_\omega\frac{\omega}{2k}P_{kk} - \beta_\omega\bar{\rho}\omega^2$$
$$+\sigma_d\frac{\bar{\rho}}{\omega}\max\left(\frac{\partial k}{\partial x_k}\frac{\partial\omega}{\partial x_k}, 0\right) \tag{43}$$

Table 2 Bounding values of the ω-equation coefficients

	σ_ω	α_ω	β_ω	σ_d
ϕ^ε	0.856	0.44	0.0828	1.712
ϕ^ω	0.5	0.5556	0.075	0

The coefficients ϕ_c=($\sigma_\omega, \alpha_\omega, \beta_\omega, \sigma_d$) in the ω-equation are also blended as

$$\phi_c = F_1\phi^\omega + (1 - F_1)\,\phi^\varepsilon \tag{44}$$

where the coefficients of ϕ^ω and ϕ^ε are given in Table 2.

2.2.3 GLVY Model

The GLVY model is the latest evolution of the GV model [11] and is fully wall-normal-free (similar to the MCL model)

$$\Pi_{ij} = \Pi_{ij}^{\mathrm{RH}} + \Pi_{ij}^{\mathrm{RI}} + \Pi_{ij}^{\mathrm{SH}} + \Pi_{ij}^{\mathrm{SI}} \tag{45}$$

where the rapid part, $\Pi_{ij}^{\mathrm{RH}} + \Pi_{ij}^{\mathrm{RI}}$, is taken directly from the GV model with slight modifications of its coefficients. The slow part does not include the anisotropy of dissipation, as it contains two novel inhomogeneous terms.

$$\Pi_{ij}^{\mathrm{RH}} = -C_\Pi^{\mathrm{RH}}\left(P_{ij} - \frac{1}{3}P_{kk}\delta_{ij}\right) \tag{46}$$

$$\Pi_{ij}^{Ri} = C_\Pi^{\mathrm{RI}}\left(\Pi_{nm}^{\mathrm{RH}} e_{I_n} e_{I_m}\delta_{ij} - \frac{3}{2}\Pi_{in}^{\mathrm{RH}} e_{I_n} e_{I_j} - \frac{3}{2}\Pi_{jn}^{\mathrm{RH}} e_{I_n} e_{I_i}\right) \tag{47}$$

$$\Pi_{ij}^{\mathrm{SH}} = -C_\Pi^{\mathrm{SH}}\bar{\rho}\varepsilon^* a_{ij} \tag{48}$$

$$\Pi_{ij}^{\mathrm{SI}} = \Pi_{ij}^{\mathrm{SI1}} + \Pi_{ij}^{\mathrm{SI2}} + \Pi_{ij}^{\mathrm{SI3}} \tag{49}$$

$$\Pi_{ij}^{\mathrm{SI1}} = C_\Pi^{\mathrm{SI1}}\frac{\bar{\rho}\varepsilon^*}{k}\left(\tilde{\mathfrak{R}}_{nm} e_{I_n} e_{I_m}\delta_{ij} - \frac{3}{2}\tilde{\mathfrak{R}}_{ni} e_{I_n} e_{I_j} - \frac{3}{2}\tilde{\mathfrak{R}}_{nj} e_{I_n} e_{I_i}\right) \tag{50}$$

$$\Pi_{ij}^{\mathrm{SI2}} = -C_\Pi^{\mathrm{SI2}}\frac{\bar{\rho}k}{\varepsilon}\frac{\partial k}{\partial x_l}\left(a_{ik}\frac{\partial\tilde{\mathfrak{R}}_{kj}}{\partial x_l} + a_{jk}\frac{\partial\tilde{\mathfrak{R}}_{ki}}{\partial x_l} - \frac{2}{3}a_{mk}\frac{\partial\tilde{\mathfrak{R}}_{km}}{\partial x_l}\delta_{ij}\right) \tag{51}$$

$$\Pi_{ij}^{\mathrm{SI3}} = C_\Pi^{\mathrm{SI3}}\left(\Pi_{nm}^{\mathrm{SI2}} e_{I_n} e_{I_m}\delta_{ij} - \frac{3}{2}\Pi_{in}^{\mathrm{SI2}} e_{I_n} e_{I_j} - \frac{3}{2}\Pi_{jn}^{\mathrm{SI2}} e_{I_n} e_{I_i}\right) \tag{52}$$

with the coefficients

$$C_{\Pi}^{\mathrm{RH}} = \min\left[1, 0.75 + 1.3\max\left(0, A - 0.55\right)\right]\left[1 - \max\left(0, 1 - Re_t/50\right)\right] \times A^{\{\max[0.25, 0.5 - 1.3\max(0, A - 0.55)]\}} \tag{53}$$

$$C_{\Pi}^{\mathrm{RI}} = \max\left(\frac{2}{3} - \frac{1}{6C_{\Pi}^{\mathrm{RH}}}, 0\right)\sqrt{\frac{\partial l_{\mathrm{TRI}}}{\partial x_l}\frac{\partial l_{\mathrm{TRI}}}{\partial x_l}}, \quad l_{\mathrm{TRI}} = \frac{l_t\left(1 - e^{-Re_t^*/30}\right)}{1 + 1.6A_2^{\max(0.6, A)}} \tag{54}$$

$$C_{\Pi}^{\mathrm{SH}} = 3.7\mathrm{AA}_2^{1/4}\left(1 - e^{-(Re_t/130)^2}\right) \tag{55}$$

$$C_{\Pi}^{\mathrm{SI1}} = -\frac{4}{9}\left(C_{\Pi}^{\mathrm{SH}} - \frac{9}{4}\right)\sqrt{\frac{\partial l_{\mathrm{TSI}}}{\partial x_l}\frac{\partial l_{\mathrm{TSI}}}{\partial x_l}}, \quad l_{\mathrm{TSI}} = \frac{l_t\left(1 - e^{-Re_t^*/30}\right)}{1 + 2.9\sqrt{A_2}} \tag{56}$$

$$C_{\Pi}^{\mathrm{SI2}} = 0.002 \tag{57}$$

$$C_{\Pi}^{\mathrm{SI3}} = 0.14\sqrt{\frac{\partial l_t^*}{\partial x_l}\frac{\partial l_t^*}{\partial x_l}} \tag{58}$$

The unit vector pointing into the direction of the non-homogeneity of the turbulence field, $\boldsymbol{e}_I$ is given by:

$$e_{I_i} = \frac{\frac{\partial l_{\mathrm{TE}}}{\partial x_i}}{\sqrt{\frac{\partial l_{\mathrm{TE}}}{\partial x_l}\frac{\partial l_{\mathrm{TE}}}{\partial x_l}}}, \quad l_{\mathrm{TE}} = \frac{l_t\left(1 - e^{-Re_t^*/30}\right)}{1 + 2\sqrt{A_2} + 2A^{16}} \tag{59}$$

The dissipation model is based on Rotta's [23] model with an appropriate damping function:

$$\varepsilon_{ij} = \frac{2}{3}\bar{\rho}\varepsilon\left(1 - f_\varepsilon\right)\delta_{ij} + \frac{\bar{\rho}\varepsilon}{k}f_\varepsilon\tilde{\mathfrak{R}}_{ij}, \quad f_\varepsilon = 1 - \left(1 - e^{-(Re_t/10)}\right)A^{\left(1 + A^2\right)} \tag{60}$$

The turbulent diffusion model in the GLVY model is based on the approximation of the triple-velocity correlation that was proposed by Hanjalic and Launder [15]:

$$D_{ij}^t = \frac{\partial}{\partial x_l}\left(-\widetilde{\bar{\rho}u_i''u_j''u_l''}\right)$$

$$-\widetilde{\bar{\rho}u_i''u_j''u_l''} = C^{su}\frac{\bar{\rho}k}{\varepsilon}\left(\tilde{\mathfrak{R}}_{im}\frac{\partial\tilde{\mathfrak{R}}_{jl}}{\partial x_m} + \tilde{\mathfrak{R}}_{jm}\frac{\partial\tilde{\mathfrak{R}}_{li}}{\partial x_m} + \tilde{\mathfrak{R}}_{lm}\frac{\partial\tilde{\mathfrak{R}}_{ij}}{\partial x_m}\right), \quad C^{su} = 0.11 \tag{61}$$

The pressure diffusion term is given as:

$$D_{ij}^P = C^{\mathrm{SP1}} \frac{\bar{\rho} k^3}{\varepsilon^3} \frac{\partial \varepsilon^*}{\partial x_i} \frac{\partial \varepsilon^*}{\partial x_j} + C^{\mathrm{SP2}} \frac{\partial}{\partial x_l} \left(\bar{\rho} \widetilde{u_m'' u_m'' u_j''} \delta_{il} + \bar{\rho} \widetilde{u_m'' u_m'' u_i''} \delta_{jl} \right)$$
$$+ C^{\mathrm{RP}} \frac{\bar{\rho} k^2}{\varepsilon^2} S_{kl} a_{lk} \frac{\partial k}{\partial x_i} \frac{\partial k}{\partial x_j} \tag{62}$$

with the coefficients

$$C^{\mathrm{SP1}} = -0.005, \quad C^{\mathrm{SP2}} = 0.022, \quad C^{\mathrm{RP}} = -0.005 \tag{63}$$

Similar to the MCL model, the GLVY model solves the homogeneous dissipation rate, based on the Launder–Sharma model:

$$\frac{\partial \bar{\rho} \varepsilon^*}{\partial t} + \frac{\partial \bar{\rho} \varepsilon^* \tilde{u}_k}{\partial x_k} = \frac{\partial}{\partial x_l} \left[\left(\bar{\mu} \delta_{kl} + C_\varepsilon \bar{\rho} \tilde{\mathfrak{R}}_{kl} \frac{k}{\varepsilon^*} \right) \frac{\partial \varepsilon^*}{\partial x_l} \right] + C_{\varepsilon 1} \frac{\varepsilon^*}{2k} P_{kk}$$
$$- C_{\varepsilon_2} \frac{\bar{\rho} \varepsilon^* \varepsilon^*}{k} + E_\varepsilon \tag{64}$$

In the original GLVY model the term E_ε is modeled as

$$E_\varepsilon = 2 \bar{\mu} C_t \frac{k^2}{\varepsilon^*} \frac{\partial^2 \tilde{u}_i}{\partial x_l x_l} \frac{\partial^2 \tilde{u}_i}{\partial x_m x_m}, \quad C_t = 0.09 e^{-\frac{3.4}{\left(1+0.02 Re_t^*\right)^2}} \tag{65}$$

This formulation does not response to shear stresses. Therefore, in the present work this term is modeled based on the more traditional term as follows:

$$E_\varepsilon = C_{\varepsilon_3} \bar{\mu} \frac{k}{\varepsilon} \tilde{\mathfrak{R}}_{kl} \frac{\partial^2 \tilde{u}_i}{\partial x_k \partial x_j} \frac{\partial^2 \tilde{u}_i}{\partial x_l \partial x_j} \tag{66}$$

The coefficients of the length-scale equation are:

$$C_\varepsilon = 0.18, \; C_{\varepsilon_1} = 1.44, \; C_{\varepsilon_2} = 1.92 \left(1 - 0.3 e^{-\left(Re_t^*\right)^2} \right), \; C_{\varepsilon_3} = 0.3 \tag{67}$$

3 Numerical Method

The governing equations are discretized using a finite difference method on a curvilinear coordinates computational mesh. The convective flux vector is computed according to the HLLC scheme [2]. The left and right states of the primitive variables vector are evaluated using a third-order biased MUSCL method. The Van Albada limiter [1] is used to suppress spurious oscillations. The diffusive flux vector is approximated by a second-order central differencing method, using a full viscous stencil. The mid-point values are approximated by simple arithmetic averaging.

An implicit time marching scheme is used to advance the discrete equations in a pseudo time step to a steady-state solution. The time-marching approach relies on a decoupled strategy, that is, the five mean-flow equations are solved separately from the seven Reynolds-stress closure equations. Moreover, the unconditionally stable scheme for the Reynolds stress model equations [21] was adopted. This scheme significantly contributes to the overall flow solver stability, resulting in a robust flow solver as if using a conventional, two-equation turbulence model, and incurs a reasonable added cost.

4 Results and Discussion

Numerical simulations of the flow about a 6:1 prolate spheroid at a Mach number of $M_\infty = 0.2$, an angle of incidence of $\alpha = 20°$, and a Reynolds number of $Re_\infty = 4.2 \times 10^6$ are conducted. Although in the present work the transition is not modeled, it should be emphasized that in the experiments [27], a trip wire was mounted on the fore-body, at $x/L \approx 0.2$ (L denotes the prolate spheroid body length and x is the axial coordinate whose origin is at the prolate spheroid nose). The trip wire position was carefully chosen to stabilize the location of transition and, consequently, the location of the separation. Based on the experimental results, the flow field is assumed to be symmetric and therefore the computational mesh is built accordingly. Moreover, the support sting was included in the computational model, extended to about 30 times of the prolate spheroid length. A fine and a coarse grid were used, where the coarse grid was generated by eliminating every second line in each direction of the fine grid. The fine grid has the dimensions of $505 \times 121 \times 229$ (streamwise $\times$ circumental $\times$ normal) with the first grid point neighboring the wall placed at a distance of 2.5×10^{-6} of the body length, from the body surface.

Using the SSG/LRR-ω model, the initial solutions of the mean-flow equations (MF) and of the turbulence model equations are uniform and are based on free-stream values. However, when using the MCL or the GLVY models, the initial solution is based on the converged solution obtained from the simulation using the SSG/LRR-ω model. This is due to the anomaly of low-Reynolds ε-based RANS turbulence model as was analyzed by Rumsey et al. [24]. The criteria of convergence was the reduction of the mean-flow residual by six orders in magnitude.

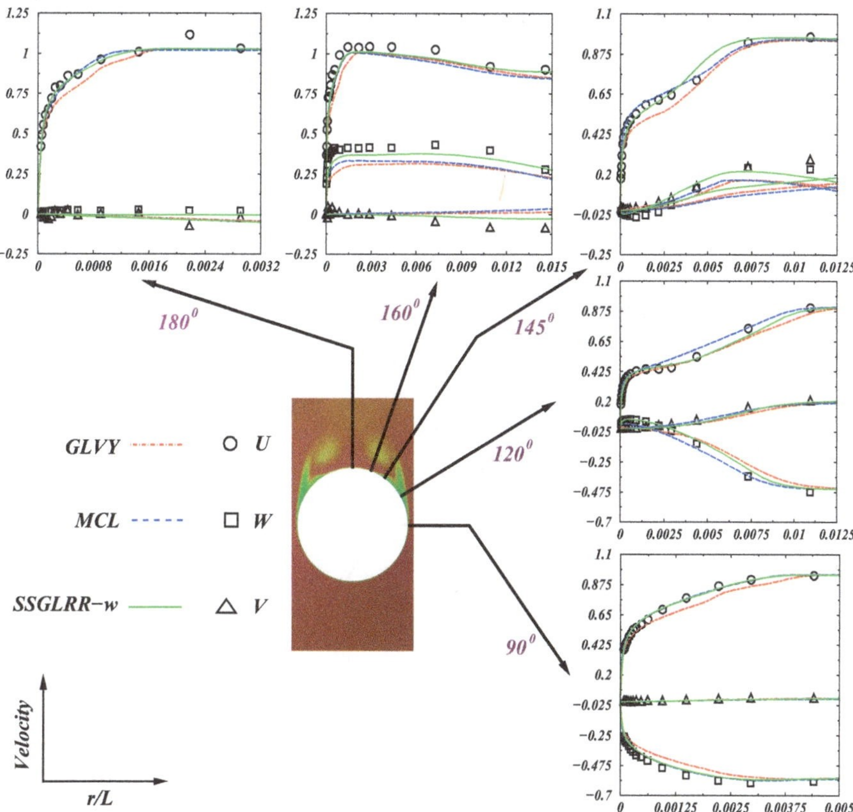

Fig. 2 Comparison between computed and measured normalized velocity profiles in the prolate spheroid body surface coordinate system at station $x/L = 0.772$ for five circumferential angles (Iso stream-wise velocity computed from the simulation using the SSG/LRR-ω model)

4.1 Velocity Profiles

Figure 2 shows a comparison of the normalized boundary layer velocity profiles, $U = \tilde{U}/\tilde{u}_\infty$, $V = \tilde{V}/\tilde{u}_\infty$, and $W = \tilde{W}/\tilde{u}_\infty$, between the measured[1] and computed results using the fine grid. The comparison is conducted for five radial rakes at the station $x/L = 0.772$. Note that the velocity components $\tilde{U}, \tilde{V}$, and $\tilde{W}$ are in body surface coordinates; $\tilde{U}$ is tangent to the hull surface and points toward the tail of the model, $\tilde{V}$ is normal to the hull surface (positive outwards), and $\tilde{W}$ is tangent to the hull surface and forms a right-handed coordinate system. At the tangential stations ϕ=90° and 180° there is an excellent agreement with the

[1]LDV and hot-wire velocity measurements of the flow about a 6:1 prolate spheroid. http://www.dept.aoe.vt.edu/~simpson/prolatespheroid/

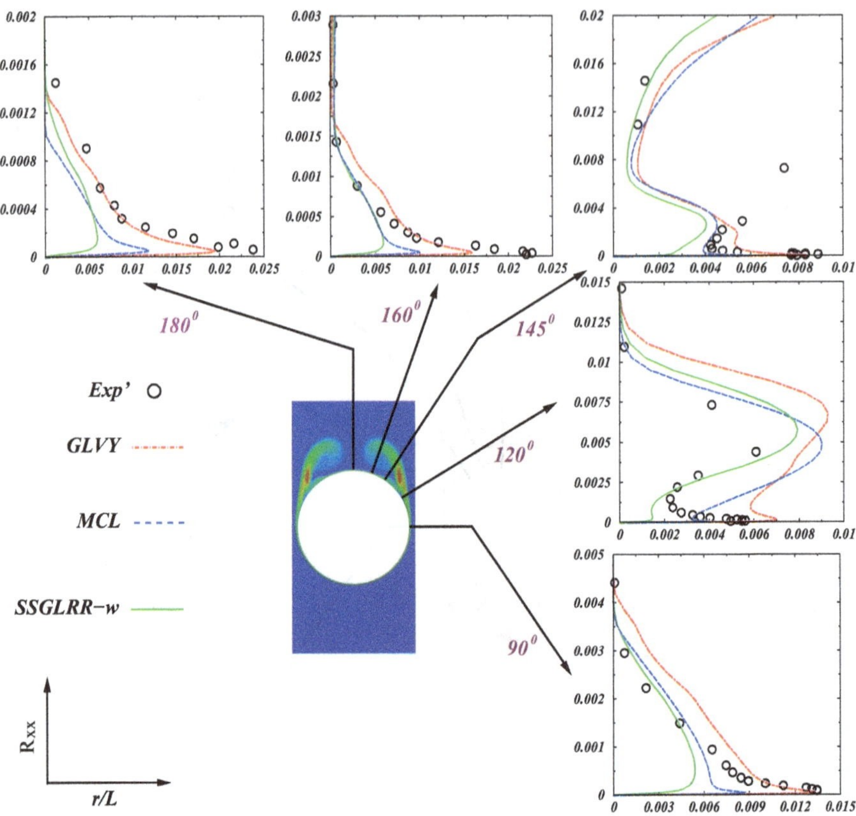

Fig. 3 Comparison between computed and measured normalized Reynolds stress component $\tilde{\mathfrak{R}}_{xx}$ profiles in the prolate spheroid body axis coordinate system at station $x/L = 0.772$ for five circumferential angles (Iso tangential Reynolds stress component computed from the simulation using the SSG/LRR-ω model)

experimental data when using the MCL and the SSG/LRR-ω models, but rather reasonable agreement when using the GLVY model. A further close examination of the other three stations shows that the SSG/LRR-ω model results in the overall best agreement with the experimental data. Especially at the tangential stations ϕ=120° where the SSG/LRR-ω accurately captures the velocity component $\tilde{W}$ even at the immediate near wall region.

4.2 Reynolds Stresses Profiles

A comparison of the normalized normal Reynolds stress components profiles, $R_{xx} = \tilde{\mathfrak{R}}_{xx}/\tilde{u}_{\infty}^2$, $R_{rr} = \tilde{\mathfrak{R}}_{rr}/\tilde{u}_{\infty}^2$, and $R_{\phi\phi} = \tilde{\mathfrak{R}}_{\phi\phi}/\tilde{u}_{\infty}^2$, and the shear component $R_{xy} = \tilde{\mathfrak{R}}_{xr}/\tilde{u}_{\infty}^2$, at five radial rakes at station $x/L = 0.772$ is shown in Figs. 3–6. Note that

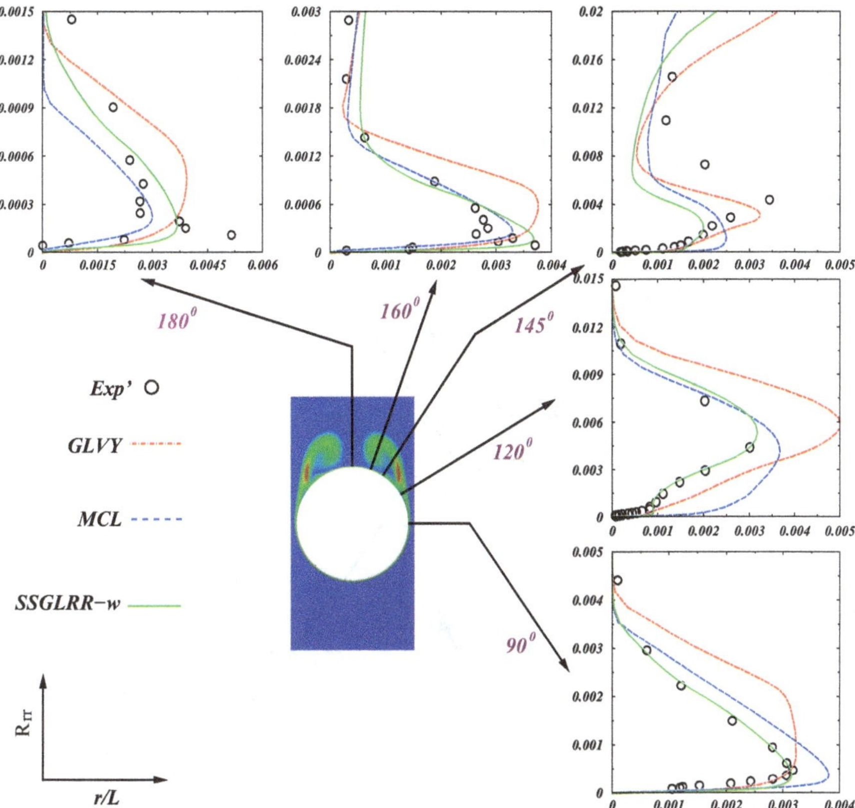

Fig. 4 Comparison between computed and measured normalized Reynolds stress component $\tilde{\mathfrak{R}}_{rr}$ profiles in the prolate spheroid body axis coordinate system at station $x/L = 0.772$ for five circumferential angles (Iso tangential Reynolds stress component computed from the simulation using the SSG/LRR-ω model)

the triad x, r, and ϕ is a body-axis coordinate system, such that x is measured from the prolate spheroid nose along the model axis. The radial distance, r, is measured perpendicular to the model axis. The azimuthal position, ϕ, is measured from the windward side of the model.

From Fig. 3 it is evident that the *GLVY* model is superior to the three other models in predicting $\tilde{\mathfrak{R}}_{xx}$. In particular, the *GLVY* model predicts the very near-wall behavior of $\tilde{\mathfrak{R}}_{xx}$ with a very good agreement with the experiment. On the other hand, the prediction of the very near wall region obtained from the SSG/LRR-ω model is in poor agreement with the experimental data.

Figures 4 and 5 clearly show that, overall, the SSG/LRR-ω model is superior to the two other models in predicting $\tilde{\mathfrak{R}}_{rr}$ and $\tilde{\mathfrak{R}}_{\phi\phi}$. Moreover, there is some similarity between the results obtained from the SSG/LRR-ω and MCL models in predicting the normal Reynolds stress $\tilde{\mathfrak{R}}_{\phi\phi}$, except at the azimuthal station ϕ=120°. At that

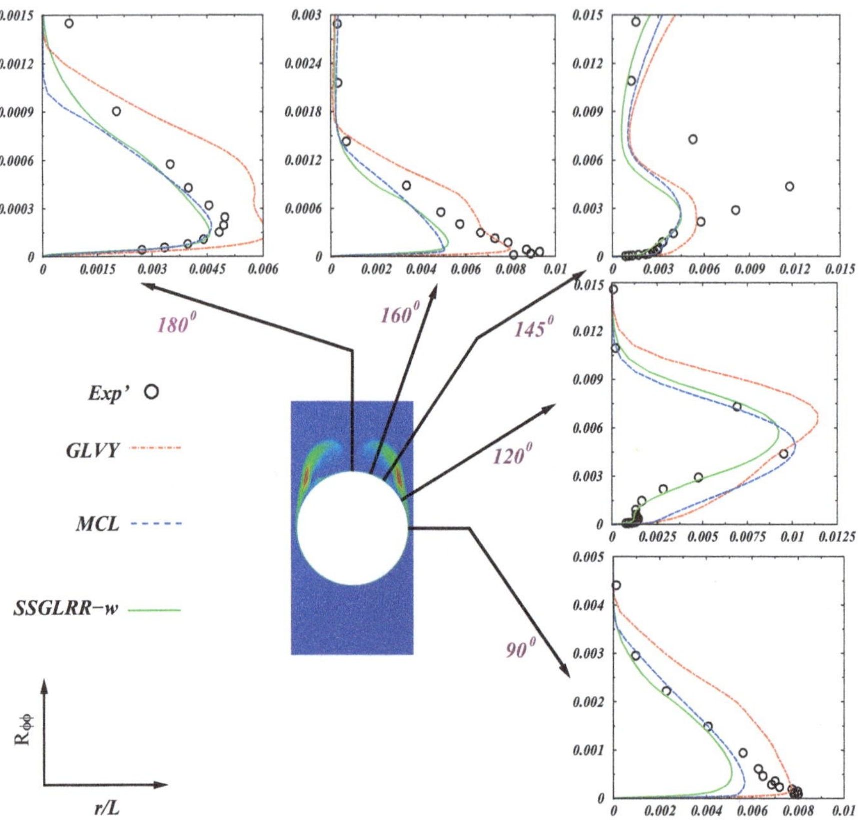

Fig. 5 Comparison between computed and measured normalized Reynolds stress component $\tilde{\mathfrak{R}}_{\phi\phi}$ profiles in the prolate spheroid body axis coordinate system at station $x/L = 0.772$ for five circumferential angles (Iso tangential Reynolds stress component computed from the simulation using the SSG/LRR-ω model)

station, the SSG/LRR-ω model has consistently a very good agreement with the experiment, except in predicting the shear stress, $\tilde{\mathfrak{R}}_{xr}$, shown in Fig. 6. In fact, none of the three models managed to predict the shear stress $\tilde{\mathfrak{R}}_{xr}$, even in a reasonable manner. Nevertheless, there is some similarity in the results obtained from the SSG/LRR-ω and the MCL models of $\tilde{\mathfrak{R}}_{xr}$ at the stations ϕ=90°, 160°, 180°.

4.3 On the Accuracy of the RSM Equations

Very often, for the purpose of numerical stability, the convective flux of RANS turbulence models and in particular of RSM is approximated using first order upwind schemes (while the convective flux of the mean-flow equation is approximated

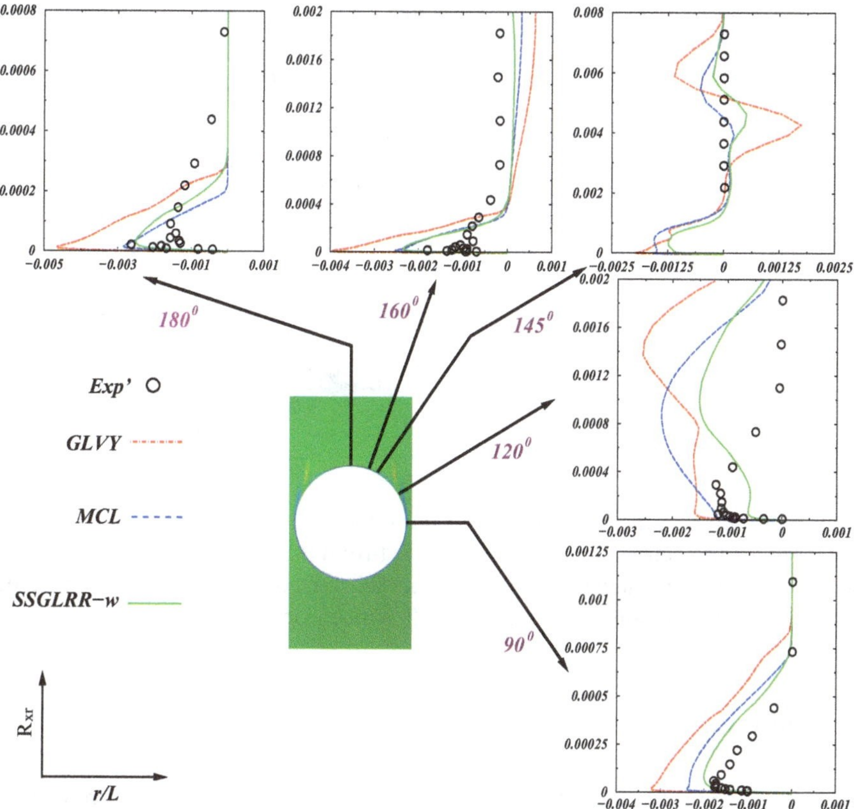

Fig. 6 Comparison between computed and measured normalized Reynolds stress component $\tilde{\mathfrak{R}}_{xr}$ profiles in the prolate spheroid body axis coordinate system at station $x/L = 0.772$ for five circumferential angles (Iso tangential Reynolds stress component computed from the simulation using the SSG/LRR-ω model)

with a higher accuracy). This approximation may be reasonable for simple attached flows. However, for a diffusion dominant region, as in separated flow regions or for unsteady flows, this may significantly degrade the solution accuracy. To examine the effect of the RSM convective flux numerical accuracy, two additional numerical simulations with the SSG/LRR-ω model using the coarse grid were conducted. In one of the two numerical simulations, the numerical scheme is left without any change i.e, third-order biased MUSCL scheme of the RSM convective flux. For the second numerical simulation, a first-order upwind scheme is used for the RSM convective flux.

Figure 7 shows a comparison of the normalized normal Reynolds stress, $\tilde{\mathfrak{R}}_{\phi\phi}$ at the azimuthal station of $\phi = 120°$. As expected, using the coarse grid, the agreement of the computed results with the experiment is degraded compared to the computed results using the fine grid. Moreover, using the coarse grid with the

Fig. 7 Comparison between computed and measured normalized Reynolds stress component $\tilde{\mathfrak{R}}_{\phi\phi}$ profiles in the prolate spheroid body axis coordinate system at the azimuthal station of $\phi = 120°$

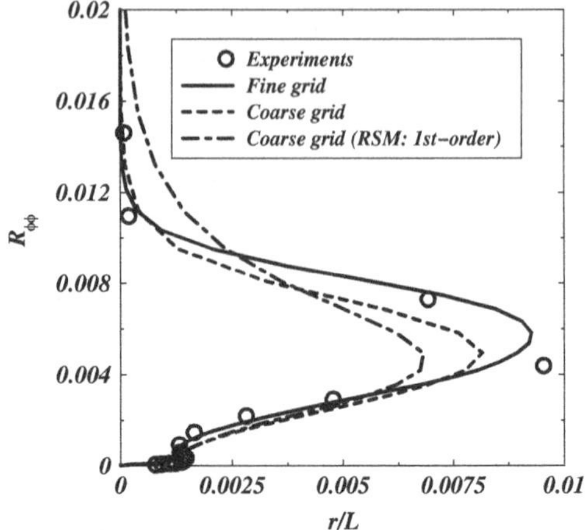

high-order scheme, the solution follows the solution obtained from the fine grid, in some sense. In contrast, the solution obtained from the coarse grid with first-order accuracy completely deviates from the solution obtained using the fine grid, i.e., the recovery of the solution away from the wall is much slower.

5 Summary

Numerical simulations of the complex flow about a 6:1 prolate spheroid were conducted using state-of-the-art Reynolds stress models. Detailed comparisons of the computed boundary layer velocity component and of the Reynolds stress components with measurement were conducted. Based on the comparisons, it is evident that overall the SSG/LRR-ω model is the best model for the prolate spheroid test case. Nevertheless, it should be noted that the better agreement of the SSG/LRR-ω model is not absolute since a large portion of laminar flow was not modeled due to the fully turbulent flow assumption undertaken in this work. Specifically, it is suspected that cross-flow transition may have a major effect on the flow characteristics.

It was demonstrated that the numerical accuracy of the Reynolds stress model alone may have a significant effect on the solution accuracy. Therefore a great caution should be taken in the a posteriori stage of a Reynolds stress model design.

References

1. van Albada GD, van Leer B, Roberts WW (1982) A comparative study of computational methods in cosmic gas dynamics. Astron Astrophys 108(1):76–84

2. Batten P, Leschziner MA, Goldberg UC (1997) Average-state Jacobians and implicit methods for compressible viscous and turbulent flows. J Comput Phys 137(1):38–78
3. Ben Nasr N, Gerolymos GA, Vallet I (2014) Low-diffusion approximate Riemann solvers for Reynolds-stress transport. J Comput Phys 268(1):186–235
4. Cécora R-D, Eisfeld B, Probst A, Crippa S, Radespiel R (2012) Differential Reynolds stress modeling for aeronautics. AIAA paper 2012-0465, 50th AIAA aerospace sciences meeting, Nashville
5. Chaouat B (2006) Reynolds stress transport modeling for high-lift airfoil flows. AIAA J 44(10):2390–2403
6. Chesnakas CJ, Taylor D, Simpson RL (1997) Detailed investigation of the three-dimensional separation about a 6:1 prolate spheroid. AIAA J 35(6):990–999
7. Constantinescu GS, Pasinato H, Wang Y-Q, Forsythe JR, Squires KD (2002) Numerical investigation of flow past a prolate spheroid. J Fluids Eng 124(4):904–910
8. Craft TJ, Launder BE (1996) A Reynolds stress closure designed for complex geometries. Int J Heat Fluid Flow 17(3):245–254
9. Eisfeld B (2004) Implementation of Reynolds stress models into the DLR-FLOWer code. IB 124-2004/31, DLR
10. Fu S (1988) Computational modelling of turbulent swirling flows with second-moment closures. University of Manchester, Institute of Science and Technology
11. Gerolymos GA, Vallet I (2001) Wall-normal-free Reynolds-stress closure for three-dimensional compressible separated flows. AIAA J 39(10):1833–1842
12. Gerolymos GA, Joly S, Mallet M, Vallet I (2010) Reynolds-stress model flow prediction in aircraft-engine intake double-S-shaped duct. J Aircr 47(4):1368–1381
13. Gerolymos GA, Lo C, Vallet I, Younis BA (2012) Term-by-term analysis of near-wall second-moment closures. AIAA J 50(12):2848–2864
14. Goody MC, Simpson RL, Chesnakas CJ (2000) Separated flow surface pressure fluctuations and pressure-velocity correlations on prolate spheroid. AIAA J 38(2):266–274
15. Hanjalić K, Launder BE (1972) A Reynolds stress model of turbulence and its application to thin shear flows. J Fluid Mech 52(4):609–638
16. Jakirlić S, Maduta R (2015) Extending the bounds of "steady" RANS closures: toward an instability-sensitive Reynolds stress model. Int J Heat Fluid Flow 51:175–194
17. Jakirlic S, Eisfeld B, Jester-Zürker R, Kroll N (2007) Near-wall, Reynolds-stress model calculations of transonic flow configurations relevant to aircraft aerodynamics. Int J Heat Fluid Flow 28:602–615
18. Kim S-E, Rhee SH, Cokljat D (2003) Application of modern turbulence models to vortical flow around a 6:1 prolate spheroid at incidence. AIAA paper 2003-0429, 41st AIAA aerospace sciences meeting and exhibit, Reno NV
19. Manceau R (2015) Recent progress in the development of the Elliptic blending Reynolds-stress model. Int J Heat Fluid Flow 51:195–220
20. Menter FR (1994) Two-equation eddy-viscosity turbulence models for engineering applications. AIAA J 32(8):1598–1605
21. Mor-Yossef Y (2014) Unconditionally stable time marching scheme for Reynolds stress models. J Comput Phys 276:635–664
22. Probst A, Radespiel R (2008) Implementation and extension of a near-wall Reynolds-stress model for application to aerodynamic flows on unstructured meshes. AIAA paper 2008-770, 46th AIAA aerospace sciences meeting and exhibit, Reno NV
23. Rotta JC (1951) Statistische theorie nichthomogener turbulenz. Z für Phys 129(6):547–572
24. Rumsey CL, Pettersson Reif BA, Gatski TB (2006) Arbitrary steady-state solutions with the k-ε model. AIAA J 44(7):1586–1592
25. Scott NW, Duque EPN (2004) Unsteady Reynolds-averaged Navier-Stokes predictions of the flow around a prolate spheroid. AIAA paper 2004-0055, 42nd AIAA aerospace sciences meeting and exhibit, Reno NV
26. Tsai C-Y, Whitney AK (1999) Numerical study of three-dimensional flow separation for a 6:1 ellipsoid. AIAA paper 99-0172, 37th AIAA aerospace sciences meeting and exhibit, Reno NV

27. Wetzel TG, Simpson RL, Chesnakas CJ (1998) Measurement of three-dimensional crossflow separation. AIAA J 36(4):557–564
28. Wikström N, Svennberg U, Alin N, Fureby C (2004) Large eddy simulation of the flow around an inclined prolate spheroid. J Turbul 5(29):1–4
29. Xiao Z, Zhang Y, Huang J, Chen H, Fu S (2007) Prediction of separation flows around a 6:1 prolate spheroid using RANS/LES hybrid approaches. Acta Mech Sin 23(4):369–382

Influence of Pressure-Strain Closure on the Prediction of Separated Flows

G.A. Gerolymos and I. Vallet

Abstract This chapter investigates the influence of the pressure-strain and pressure-diffusion closures in differential RSMs (Reynolds-stress models). The development of different modelling strategies for the terms containing the fluctuating pressure p' in the Reynolds-stress transport equations is reviewed. The retained model is then assessed by comparison with selected test-cases from the NASA Turbmodels website (http://turbmodels.larc.nasa.gov/), *viz* 2-D ZPG (zero-pressure-gradient) flat-plate boundary-layer, 2-D airfoil (NACA 4412) trailing-edge separation, 2-D convex curvature boundary-layer, and 3-D supersonic square duct flow. The influence of different modelling choices is illustrated by comparison with test-model variants, with different coefficient functions in selected terms. Perspectives for the development of improved differential Reynolds-stress models are then discussed, in particular with regard to the development of a new 12-equation r_{ij}–ε_{ij} family of models, incorporating transport equations for the different components of the dissipation tensor.

1 Differential Reynolds-Stress Models

Computationally efficient RANS calculations are necessary in the aerospace design process [59], where a large number of simulations must be performed. Flow separation, that often dominates complex practical flows, even at nominal operating conditions [24] presents marked anisotropy [51] and strong hysteresis [17] phenomena which require advanced turbulence closures to achieve acceptable accuracy. Differential Reynolds-stress models (RSMs) which directly include several important mechanisms (anisotropy, convective history, streamline curvature, redistribution, Coriolis effects) in the exact equations that are modelled are increasingly considered as a promising practical alternative [32] to the 2-equation closures that have widely dominated RANS CFD in the past two decades [6], especially with the availability of efficient and robust low-diffusion solvers [1] for Reynolds-stress

G.A. Gerolymos (✉) • I. Vallet
Sorbonne Universités, Université Pierre-et-Marie-Curie, 75005 Paris, France
e-mail: georges.gerolymos@upmc.fr; isabelle.vallet@upmc.fr

B. Eisfeld (ed.), *Differential Reynolds Stress Modeling for Separating Flows in Industrial Aerodynamics*, Springer Tracts in Mechanical Engineering, DOI 10.1007/978-3-319-15639-2_4

transport. Furthermore, RSMs, because of their inherent anisotropy resolving capability, are a natural choice for 3-D complex flows where secondary vorticity is important [28].

The paper discusses the performance of the GLVY [26] 7-equation $r_{ij} - \varepsilon^*$ wall-normal-free RSM, especially with reference to selected test-cases of the NASA Turbmodels project [45, 46]. The model performance is also compared with results from other models, both 2-equation [35] and 7-equation [9, 20, 56], to put into perspective the influence of different modelling choices for the pressure terms. Perspectives in differential Reynolds-stress modelling are discussed.

2 Flow Model and Solver

2.1 Background

RSMs solve six equations [43] for Reynolds-stress transport (RST) along with a scalar scale-determining [62] equation. The pressure terms in the RST equations [3], and the associated slow and rapid redistribution concepts [39, 43], are the most important improvement compared to eddy-viscosity models [63]. The initial concepts of return-to-isotropy for the slow part [43] and isotropization-of-production for the rapid part [34, 39], along with the wall-echo concept [29, 49] led to the first working models for wall-bounded flows [29]. The idea of tensorial representation for modelling the pressure-terms, present in [34], was formalized by Lumley [37], who introduced the anisotropy invariants and the flatness parameter in turbulence modelling, arguing in particular that model coefficients should not be constants, but functions of the invariants [25, 55].

Launder and Shima [36] optimized the model coefficients for near equilibrium wall-bounded flows. The Launder-Shima RSM, which adopts the tensorial representations of quasi-linear return-to-isotropy [43] and isotropization-of-production models [39], contains wall-topology-dependent terms and underestimates separation [8]. Gerolymos and Vallet [9] developed a fully wall-normal-free (wall-topology-independent) model (GV RSM) based on a modelled unit-vector pointing in the direction of inhomogeneity, with enhanced capability to predict separation by the particular functional dependence of the rapid redistribution isotropization-of-production model coefficient [9, Fig. 4, p. 1837]. The choice of quasi-linear return-to-isotropy [43] and isotropization-of-production models [39], along with wall-echo terms (which may be wall-topology-independent) generally provides robust models [32]. There has been some debate [38] on the correctness of the wall-echo concept, but it was shown recently [27] that this approach is consistent with DNS data provided it is applied to the full velocity/pressure-gradient tensor Π_{ij}.

The recently developed GLVY RSM [26] attempts to improve upon the GV RSM [9], by including specific modelling for pressure-diffusion, nonlinear inhomogeneous terms in the slow redistribution model, and a separate model for the dissipation-rate-tensor anisotropy (which is often modelled together with the slow part of redistribution [36] following Lumley's [37] suggestion). The GLVY RSM [26] slightly improves upon the GV RSM [9] in the reattachment and relaxation region, and also has a different apparent transition behaviour [15].

2.2 *GLVY RSM [26]*

Details on the development of the RSMs used in the present work can be found in the original papers [9, 20, 26]. They are summarized below for completeness, in a common representation which highlights differences in the closure choices between different models. The model equations for r_{ij} and ε^* read

$$\underbrace{\frac{\partial}{\partial t}(\bar{\rho}r_{ij})+\frac{\partial}{\partial x_\ell}(\bar{\rho}r_{ij}\tilde{u}_\ell)}_{C_{ij}}=\underbrace{-\bar{\rho}r_{i\ell}\frac{\partial\tilde{u}_j}{\partial x_\ell}-\bar{\rho}r_{j\ell}\frac{\partial\tilde{u}_i}{\partial x_\ell}}_{P_{ij}}+\underbrace{\frac{\partial}{\partial x_\ell}\left(\breve{\mu}\frac{\partial r_{ij}}{\partial x_\ell}\right)}_{d_{ij}^{(\mu)}}$$

$$+d_{ij}^{(u)}+\Pi_{ij}-\bar{\rho}\varepsilon_{ij}+K_{ij} \tag{1a}$$

$$\frac{\partial\bar{\rho}\varepsilon^*}{\partial t}+\frac{\partial(\bar{\rho}\varepsilon^*\tilde{u}_\ell)}{\partial x_\ell}=\frac{\partial}{\partial x_\ell}\left[C_\varepsilon\frac{\mathrm{k}}{\varepsilon^*}\bar{\rho}r_{m\ell}\frac{\partial\varepsilon^*}{\partial x_m}+\breve{\mu}\frac{\partial\varepsilon^*}{\partial x_\ell}\right]$$

$$+C_{\varepsilon_1}P_\mathrm{k}\frac{\varepsilon^*}{\mathrm{k}}-C_{\varepsilon_2}\bar{\rho}\frac{\varepsilon^{*2}}{\mathrm{k}}+2\breve{\mu}C_\mu\frac{\mathrm{k}^2}{\varepsilon^*}\frac{\partial^2\tilde{u}_i}{\partial x_\ell\partial x_\ell}\frac{\partial^2\tilde{u}_i}{\partial x_m\partial x_m} \tag{1b}$$

$$P_\mathrm{k}:=\tfrac{1}{2}P_{\ell\ell}\ ;\ C_\varepsilon=0.18\ ;\ C_{\varepsilon 1}=1.44 \tag{1c}$$

$$C_{\varepsilon 2}=1.92(1-0.3\mathrm{e}^{-Re_\mathrm{T}^{*2}})\ ;\ C_\mu=0.09\mathrm{e}^{-\frac{3.4}{(1+0.02Re_\mathrm{T}^*)^2}} \tag{1d}$$

$$r_{ij}:=\frac{1}{\bar{\rho}}\overline{\rho u_i''u_j''}\ ;\ \mathrm{k}:=\tfrac{1}{2}r_{\ell\ell}\ ;\ Re_\mathrm{T}^*:=\frac{\mathrm{k}^2}{\breve{\nu}\varepsilon^*}\ ;\ \breve{\mu}:=\mu_\mathrm{Sutherland}(\tilde{T})\ ;\ \breve{\nu}:=\frac{\breve{\mu}}{\bar{\rho}} \tag{1e}$$

In (1), t is the time, $u_i\in\{u,v,w\}$ are the velocity components in the Cartesian reference-frame $x_i\in\{x,y,z\}$, ρ is the density, p is the pressure, r_{ij} (1e) are the Favre-averaged Reynolds-stresses, $\breve{\mu}$ is the dynamic viscosity evaluated from Sutherland's law [57, (6), p. 528] at mean temperature $\tilde{T}$, $\breve{\nu}$ is the kinematic viscosity at mean-flow conditions $\overline{(\cdot)}$ denotes Reynolds averaging, $(\cdot)'$ are Reynolds fluctuations, $\tilde{(\cdot)}$ denotes Favre averaging, $(\cdot)''$ are Favre fluctuations, and $\breve{(\cdot)}$ denotes a function of averaged quantities that cannot be identified with a Reynolds or a Favre average [7, 16]. The modified dissipation-rate ε^* [8, 35] is used as scale-determining variable, and follows the modelled equation (1b), where P_k is the production-rate of turbulent kinetic energy k and Re_T^* is the turbulent Reynolds-number.

Convection C_{ij}, production P_{ij} and viscous diffusion $d_{ij}^{(\mu)}$ in (1a) are computable terms, whereas diffusion by the fluctuating velocity field $d_{ij}^{(u)}$, the velocity/pressure-gradient correlation Π_{ij}, the dissipation-tensor ε_{ij} and the fluctuating-density terms K_{ij} require closure.

The fluctuating-density terms K_{ij} were neglected, turbulent diffusion $d_{ij}^{(u)}$ was modelled by the Hanjalić-Launder [31] closure, and a wall-normal-free algebraic closure was developed [26] for ε_{ij} based on Rotta's [43]

$$K_{ij} = 0 \tag{2a}$$

$$d_{ij}^{(u)} := \frac{\partial}{\partial x_\ell}\left(-\overline{\rho u_i'' u_j'' u_\ell''}\right) \tag{2b}$$

$$-\overline{\rho u_i'' u_j'' u_k''} = C^{(\mathrm{Su})}\frac{\mathrm{k}}{\varepsilon}\left(\bar{\rho} r_{im}\frac{\partial r_{jk}}{\partial x_m} + \bar{\rho} r_{jm}\frac{\partial r_{ki}}{\partial x_m} + \bar{\rho} r_{km}\frac{\partial r_{ij}}{\partial x_m}\right) \;;\; C^{(\mathrm{Su})} = 0.11 \tag{2c}$$

$$\bar{\rho}\varepsilon_{ij} = \tfrac{2}{3}\bar{\rho}\varepsilon\left(1 - f_\varepsilon\right)\delta_{ij} + f_\varepsilon\frac{\varepsilon}{\mathrm{k}}\bar{\rho} r_{ij} \;;\; \varepsilon^* := \varepsilon - 2\breve{\nu}\frac{\partial\sqrt{\mathrm{k}}}{\partial x_\ell}\frac{\partial\sqrt{\mathrm{k}}}{\partial x_\ell} \tag{2d}$$

$$f_\varepsilon = 1 - A^{[1+A^2]}\left[1 - \mathrm{e}^{-\frac{Re_\mathrm{T}}{10}}\right] \;;\; Re_\mathrm{T} := \frac{\mathrm{k}^2}{\breve{\nu}\varepsilon} \tag{2e}$$

$$a_{ij} := \frac{r_{ij}}{\mathrm{k}} - \tfrac{2}{3}\delta_{ij} \;;\; A_2 := a_{ik}a_{ki} \;;\; A_3 := a_{ik}a_{kj}a_{ji} \;;\; A := 1 - \tfrac{9}{8}(A_2 - A_3) \tag{2f}$$

In (2), $\varepsilon := \frac{1}{2}\varepsilon_{\ell\ell}$ is the dissipation-rate of the turbulent kinetic energy, a_{ij} is the Reynolds-stress-anisotropy tensor [36], with invariants A_2 and A_3 [36], A is Lumley's flatness parameter [37], and Re_T is the turbulent Reynolds number. Finally, the velocity/pressure-gradient tensor Π_{ij} is modelled as

$$\Pi_{ij} = \underbrace{\overbrace{\phi_{ij}^{(\mathrm{RH})} + \phi_{ij}^{(\mathrm{RI})}}^{\phi_{ij}^{(\mathrm{R})}} + \overbrace{\phi_{ij}^{(\mathrm{SH})} + \phi_{ij}^{(\mathrm{SI})}}^{\phi_{ij}^{(\mathrm{S})}}}_{\phi_{ij}} + \tfrac{2}{3}\phi_p\delta_{ij} + d_{ij}^{(p)} \tag{3a}$$

$$\phi_p = 0 \tag{3b}$$

$$d_{ij}^{(p)} = C^{(\mathrm{Sp1})}\bar{\rho}\frac{\mathrm{k}^3}{\varepsilon^3}\frac{\partial\varepsilon^*}{\partial x_i}\frac{\partial\varepsilon^*}{\partial x_j} + \frac{\partial}{\partial x_\ell}\left[C^{(\mathrm{Sp2})}\left(\overline{\rho u_m'' u_m'' u_j''}\delta_{i\ell} + \overline{\rho u_m'' u_m'' u_i''}\delta_{j\ell}\right)\right] + C^{(\mathrm{Rp})}\bar{\rho}\frac{\mathrm{k}^2}{\varepsilon^2}\breve{S}_{k\ell}a_{\ell k}\frac{\partial\mathrm{k}}{\partial x_i}\frac{\partial\mathrm{k}}{\partial x_j} \tag{3c}$$

$$C^{(\mathrm{Sp1})} = -0.005 \;;\; C^{(\mathrm{Sp2})} = +0.022 \;;\; C^{(\mathrm{Rp})} = -0.005 \tag{3d}$$

$$\breve{S}_{ij} := \frac{1}{2}\left(\frac{\partial \tilde{u}_i}{\partial x_j} + \frac{\partial \tilde{u}_j}{\partial x_i}\right) \tag{3e}$$

$$\begin{aligned}\phi_{ij}^{(\mathrm{R})} =& \underbrace{-C_\phi^{(\mathrm{RH})}\left(P_{ij} - \tfrac{1}{3}\delta_{ij}P_{mm}\right)}_{\phi_{ij}^{(\mathrm{RH})}} \\ & \underbrace{+C_\phi^{(\mathrm{RI})}\left[\phi_{nm}^{(\mathrm{RH})} e_{\mathrm{I}_n} e_{\mathrm{I}_m}\delta_{ij} - \tfrac{3}{2}\phi_{in}^{(\mathrm{RH})} e_{\mathrm{I}_n} e_{\mathrm{I}_j} - \tfrac{3}{2}\phi_{jn}^{(\mathrm{RH})} e_{\mathrm{I}_n} e_{\mathrm{I}_i}\right]}_{\phi_{ij}^{(\mathrm{RI})}}\end{aligned} \tag{3f}$$

$$\begin{aligned}\phi_{ij}^{(\mathrm{S})} =& \underbrace{-C_\phi^{(\mathrm{SH1})}\bar{\rho}\varepsilon^* a_{ij}}_{\phi_{ij}^{(\mathrm{SH1})}} \underbrace{+C_\phi^{(\mathrm{SI1})}\frac{\varepsilon^*}{\mathrm{k}}\left[\bar{\rho}r_{nm}e_{\mathrm{I}_n}e_{\mathrm{I}_m}\delta_{ij} - \tfrac{3}{2}\bar{\rho}r_{ni}e_{\mathrm{I}_n}e_{\mathrm{I}_j} - \tfrac{3}{2}\bar{\rho}r_{nj}e_{\mathrm{I}_n}e_{\mathrm{I}_i}\right]}_{\phi_{ij}^{(\mathrm{SI1})}} \\ & \underbrace{-C_\phi^{(\mathrm{SI2})}\bar{\rho}\frac{\mathrm{k}}{\varepsilon}\frac{\partial \mathrm{k}}{\partial x_\ell}\left[a_{ik}\frac{\partial r_{kj}}{\partial x_\ell} + a_{jk}\frac{\partial r_{ki}}{\partial x_\ell} - \tfrac{2}{3}\delta_{ij}a_{mk}\frac{\partial r_{km}}{\partial x_\ell}\right]}_{\phi_{ij}^{(\mathrm{SI2})}} \\ & \underbrace{+C_\phi^{(\mathrm{SI3})}\left[\phi_{nm}^{(\mathrm{SI2})} e_{\mathrm{I}_n} e_{\mathrm{I}_m}\delta_{ij} - \tfrac{3}{2}\phi_{in}^{(\mathrm{SI2})} e_{\mathrm{I}_n} e_{\mathrm{I}_j} - \tfrac{3}{2}\phi_{jn}^{(\mathrm{SI2})} e_{\mathrm{I}_n} e_{\mathrm{I}_i}\right]}_{\phi_{ij}^{(\mathrm{SI3})}}\end{aligned} \tag{3g}$$

$$e_{\mathrm{I}_i} := \frac{\dfrac{\partial}{\partial x_i}\left(\dfrac{\ell_\mathrm{T}[1-\mathrm{e}^{-\frac{Re_\mathrm{T}^*}{30}}]}{1+2\sqrt{A_2}+2A^{16}}\right)}{\sqrt{\dfrac{\partial}{\partial x_\ell}\left(\dfrac{\ell_\mathrm{T}[1-\mathrm{e}^{-\frac{Re_\mathrm{T}^*}{30}}]}{1+2\sqrt{A_2}+2A^{16}}\right)\dfrac{\partial}{\partial x_\ell}\left(\dfrac{\ell_\mathrm{T}[1-\mathrm{e}^{-\frac{Re_\mathrm{T}^*}{30}}]}{1+2\sqrt{A_2}+2A^{16}}\right)}} \;;\; \ell_\mathrm{T} := \frac{\mathrm{k}^{\frac{3}{2}}}{\varepsilon} \tag{3h}$$

$$C_\phi^{(\mathrm{RH})} = \left(1 - \max\left(0, 1 - \tfrac{1}{50}Re_\mathrm{T}\right)\right) \times \begin{cases} 0.75\sqrt{A} & 0 \le A < 0.55 \\ (0.75+1.3(A-0.55))A^{0.5-1.3(A-0.55)} & 0.55 \le A < 0.55 + \frac{0.25}{1.3} \\ A^{\frac{1}{4}} & 0.55 + \frac{0.25}{1.3} \le A \le 1 \end{cases} \tag{3i}$$

$$C_\phi^{(\mathrm{RI})} = \max\left[\frac{2}{3} - \frac{1}{6C_\phi^{(\mathrm{RH})}}, 0\right]\left|\mathrm{grad}\left(\frac{\ell_\mathrm{T}[1-\mathrm{e}^{-\frac{Re_\mathrm{T}^*}{30}}]}{1+1.6A_2^{\max(0.6,A)}}\right)\right| \tag{3j}$$

$$C_\phi^{(\mathrm{SH1})} = 3.7AA_2^{\frac{1}{4}}\left[1 - \mathrm{e}^{-\left(\frac{Re_\mathrm{T}}{130}\right)^2}\right] \tag{3k}$$

$$C_\phi^{(\text{SI1})} = \left[-\tfrac{4}{9}\left(C_\phi^{(\text{SH1})} - \tfrac{9}{4}\right)\right] \left|\text{grad}\left(\frac{\ell_\text{T}[1-\text{e}^{-\frac{Re_\text{T}^*}{30}}]}{1+2.9\sqrt{A_2}}\right)\right| \tag{3l}$$

$$C_\phi^{(\text{SI2})} = 0.002 \;\; ; \;\; C_\phi^{(\text{SI3})} = 0.14\sqrt{\frac{\partial \ell_\text{T}^*}{\partial x_\ell}\frac{\partial \ell_\text{T}^*}{\partial x_\ell}} \;\; ; \;\; \ell_\text{T}^* := \frac{\text{k}^{\frac{3}{2}}}{\varepsilon^*} \tag{3m}$$

In (3), ϕ_{ij} denotes the redistribution tensor, $d_{ij}^{(p)}$ denotes pressure-diffusion, ϕ_p is the pressure-dilatation correlation which was neglected, $\breve{S}_{ij}$ is the deformation-rate tensor of the mean-velocity field, ℓ_T (ℓ_T^*) is the turbulent length scale (defined using either ε or ε^*), the superscripts S and R denote slow and rapid terms [30], the superscripts H and I denote homogeneous and inhomogeneous terms [27], and the unit-vector $\mathbf{e}_\text{I}$ was modelled [9] to point in the main direction of turbulence-inhomogeneity [20]. Notice that the expression for $C_\phi^{(\text{RH})}$ can also be written [9, 26] in an ifless form.[1] The closure (3) must be considered as a whole, and inhomogeneous terms should be kept when computing free shear flows, because they are also active at the shear-layer edge and in regions of recirculating flow, away from or even in absence of solid walls.

To put the model's performance into perspective, computations (Sect. 3) were also run with the linear eddy-viscosity 2-equation Launder-Sharma k–ε [35] model (hereafter LS k–ε) and with the WNF–LSS RSM [20]. This RSM differs from the GLVY RSM [26], principally, in the coefficient $C_\phi^{(\text{RH})}$ (which in the WNF–LSS adopts the Launder-Shima [36] proposal of $0.75\sqrt{A}$) and the absence of pressure-diffusion.

2.3 *Flow Solver*

Computations were performed using WENO3 [23] reconstruction of the primitive variables, both mean-flow and turbulent, a HLLC$_\text{h}$ approximate Riemann solver [1], and implicit multigrid dual-time-stepping pseudo-time-marching integration [11, 14]. This methodology is implemented in the open source software `aerodynamics` [13] with which the present results were obtained.

3 Assessment

A successful turbulence closure is one that predicts with reasonable accuracy the largest possible selection of flows, as opposed to a model that is fine-tuned for a specific class of flows. For this reason, the GLVY RSM (like its predecessor GV) has been assessed, indeed calibrated, against experimental (and/or DNS) data for a wide

[1] $C_\phi^{(\text{RH})} = \min[1, 0.75+1.3\max[0, A-0.55]]\; A^{[\max(0.25, 0.5-1.3\max[0,A-0.55])]}[1-\max(0, 1-\frac{Re_\text{T}}{50})]$.

variety of flows, including fully developed turbulent plane channel flow [20, 26, 47, 56], ZPG flat-plate boundary-layer flow [9, 26] including the effects of external flow turbulence [22] and the apparent transition behaviour of the model [15], flow over airfoils [1, 15], 2-D and 3-D shock-wave/boundary-layer interactions [9, 20–22, 26, 47, 57] for shock-wave Mach-numbers $\breve{M}_{SW} \in [1.1, 5]$, 2-D and 3-D separated flow in diffusers and S-shaped ducts [24, 26] secondary flows in 3-D ducts [20, 56], and complex turbomachinery flows [10, 12, 18, 19]. In the present work, the GLVY RSM is assessed against experimental data for four new test-cases from the NASA Turbmodels website [45].

3.1 ZPG Flat-Plate Boundary-Layer

This basic validation test-case [45] studies the initial (momentum-thickness Reynolds number $Re_\theta \lessapprox 14,000$) development of the ZPG boundary-layer over a flat plate of length $x_{TE} - x_{LE} = 2$ m. Computations were run with the GLVY RSM on the 545×385 grid (Fig. 1) [45]. The computational inflow is located at $x = -\frac{1}{3}$ m upstream of the plate's leading-edge ($x_{LE} = 0$), and the wall-normal height of the computational domain above the plate is $L_y = 1$ m. At inflow, uniform total pressure $p_{t_\infty} = 101,325$ Pa and total temperature $T_{t_\infty} = 302.4$ K, with turbulence intensity $T_{u_\infty} = 0.5\,\%$ and length scale $\ell_{T_\infty} = 5$ mm, were applied. At outflow ($x = x_{TE}$) and at the y-wise upper boundary ($y = L_y$), a constant pressure $p_\infty \approxeq 98,538.34$ Pa outflow condition was applied. On the plate ($x \geq 0,\ y = 0$) a no-slip adiabatic wall condition was applied, while upstream of the leading-edge ($x < x_{LE},\ y = 0$) a symmetry condition was applied. The resulting flow Mach number is $M_\infty = 0.2$ with unit-Reynolds-number $Re_1 \approxeq 4.3 \times 10^6$.

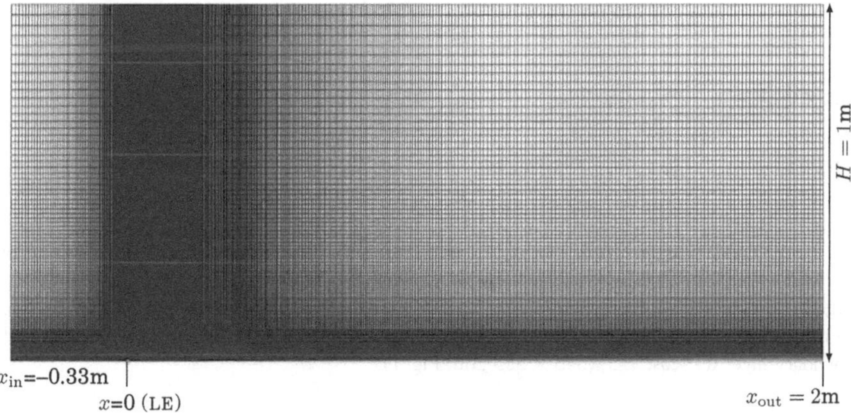

Fig. 1 Computational grid (545×385 points [45]) for the ZPG flat-plate boundary-layer test-case

Computations were run both with apparent [44] model transition and with boundary-layer tripping [15] (tripping volume at $0 \leq x \leq 0.005\,\text{m} = 5\,\text{mm}$ and $0 \leq y \leq 0.00027\,\text{m} = 270\,\mu\text{m}$ with trip intensity of $T_{u_{\text{TRIP}}} = 0.4 = 40\,\%$). With the exception of the precise location of transition, both computations, when $Re_\theta(x)$ is used to identify the x-wise evolution of the flow, yield practically identical results for $Re_\theta \gtrapprox 1{,}000$. Regarding the evolution of c_f *vs* Re_θ (Fig. 2), there is considerable scatter between different semi-empirical correlations [45], in the considered range ($Re_\theta \in [4{,}000, 13{,}000]$; gray zone, Fig. 2). The reason for this is that older experimental data for c_f were generally based on specific assumptions on the constants of the logarithmic law, which have substantial influence on the evaluation of c_f [61]. Recent direct measurement of c_f in this range [41], using oil-film interferometry [40], are in very satisfactory agreement with even more recent DNS data [48, 50] of spatially evolving boundary-layers, and with indirect (Clauser-chart) measurements [52, 58], using the different values of log-law constants ($\kappa_{\text{VK}} = 0.384$, $B_{\text{loglaw}} = 4.127$ in [58] and $\kappa_{\text{VK}} = 0.41$, $B_{\text{loglaw}} = 5.2$ in [52][2]). It appears from these recent data that the older correlations (gray zone, Fig. 2) somehow overestimate c_f in the range of interest ($Re_\theta \in [4{,}000, 13{,}000]$). Results computed using the GLVY RSM are in very satisfactory agreement with these more recent data [41, 48, 50, 52, 58], at the lower limit of the correlations scatter (Fig. 2). Regarding

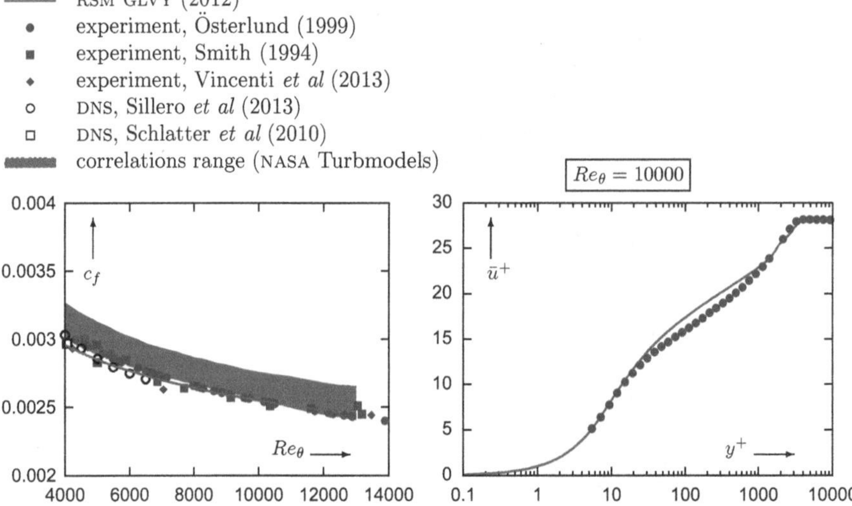

Fig. 2 Comparison of experimental [41, 52, 58] and DNS [48, 50] data of skin-friction $c_f(Re_\theta)$ and logarithmic law $\bar{u}^+(y^+)$ with computation using the GLVY RSM [26] for the ZPG flat-plate boundary-layer test-case [45] (545 × 385 grid; Fig. 1)

[2] these seemingly very different sets of log-law constants yield values of $\kappa_{\text{VK}}^{-1} \ln y^+ + B_{\text{loglaw}}$ that are quite similar in the range $y^+ \in [100, 1{,}000]$

the mean-flow velocity profile (Fig. 2), the GLVY RSM predictions are in very good agreement with experimental data, in the linear and buffer zones ($y^+ \lessapprox 30$ [63, Fig. 1.7, p. 17]) and in the wake ($y^+ \gtrapprox 1{,}000$ for the range of Reynolds numbers studied [63, Fig. 1.7, p. 17]), but the logarithmic law is slightly overestimated. The very satisfactory prediction of the linear and buffer zones implies a satisfactory prediction of skin-friction, whereas the wake, in high Reynolds-number flows, corresponds to the largest (in physical space) part of the boundary-layer. The log-law, in the GLVY RSM, was calibrated with reference to the Klebanoff's [33] measurements, as re-interpreted by So et al. [54] ($\kappa_{\text{VK}} = 0.43$, $B_{\text{loglaw}} = 5.34$ [26, Fig. 8, p. 2857]; these values for the log-law constants agree well with the previous two sets in the range $30 \leq y^+ \leq 700$). However, the calibration was done using a compressible solver [11] at very low Mach number ($M_\infty \approxeq 0.05$ in Klebanoff's [33] experiment) and this may have introduced some numerical bias.

3.2 Convex Curvature Boundary-Layer

Experimental data for this 2-D curved duct (Fig. 4) with an angle of 30 degrees [45] were obtained by Smits et al. [53], and are available in [45]. Computations were run on the fine $1{,}025 \times 385$ grid (Fig. 3) of [45]. At inflow, total pressure $p_{t_{\text{CL}_i}} = 101{,}325$ Pa and total temperature $T_{t_{\text{CL}_i}} = 293.832$ K, with turbulence intensity $T_{u_{\text{CL}_i}} = 1\,\%$ and length scale $\ell_{\text{T}_{\text{CL}_i}} = 125$ mm (which is roughly the duct's height), were applied at the centreline. The initial inflow boundary-layer thickness was $\delta_i = 0.5$ mm with $\Pi_{\text{Coles}_i} = 0$ [20]. Computations with zero inlet

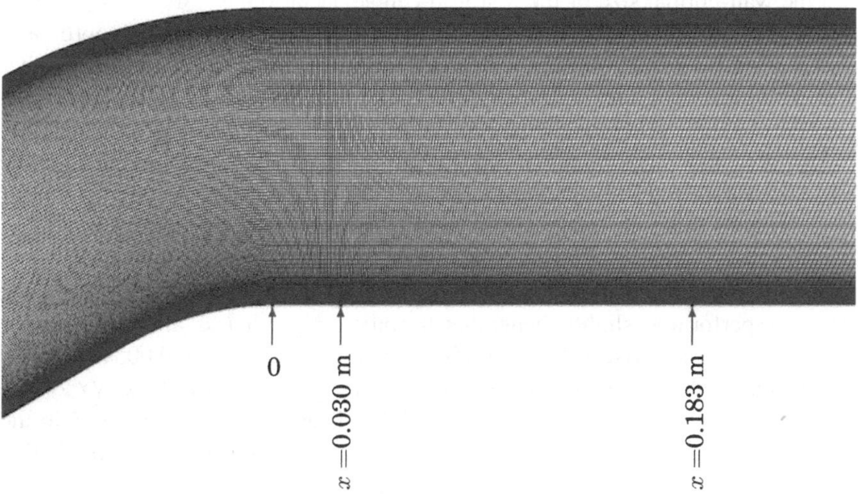

Fig. 3 View of the $1{,}025 \times 385$ points grid [45] in the neighbourhood of the Smits et al. [53] 30 deg bend

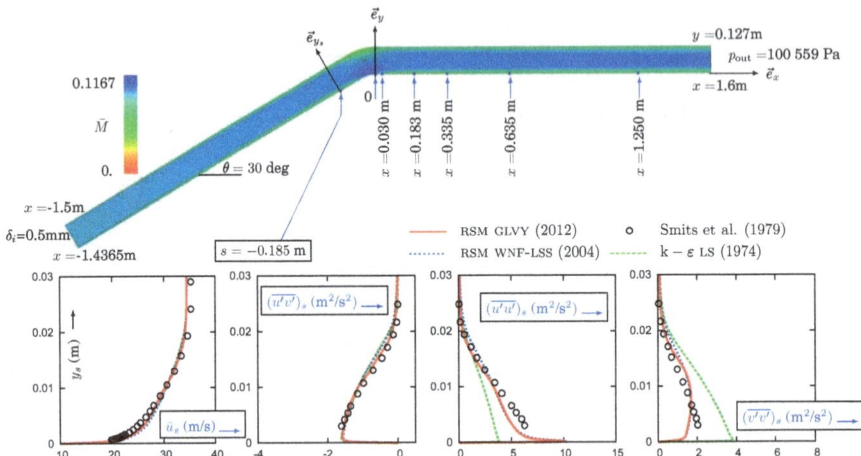

Fig. 4 Comparison of measured profiles of wall-parallel velocity $\bar{u}_s$ and Reynolds-stresses ($\overline{u_s'^2}$, $\overline{v_s'^2}$, $\overline{u_s'v_s'}$) in the wall-aligned frame, just upstream of the bend inlet (s $=$ -0.185 m; x $=$ -0.166124 m), on the convex side (*lower wall*) of the Smits et al. [53] 30 deg bend, with computations using the GLVY [26] and WNF–LSS [20] RSMs and the linear LS k–ε [35] model ($1,025 \times 385$ grid [45]; level plots of Mach number $\breve{M}$ using the GLVY RSM)

boundary-layer thickness were also run with very similar results (in this case, however, it was found necessary to apply a turbulence intensity of 2 % at inflow, to avoid relaminarization of the lower-wall boundary-layer). At outflow (Fig. 4) a constant pressure $p_\infty \approxeq 100,559$ Pa condition was applied and the walls were considered adiabatic. For the corresponding flow conditions, on the $1,025\times385$ grid [45] the wall-normal size of the first grid-cell adjacent to the wall is $\Delta n_w^+ \approxeq 0.1$, with a peak at 0.13 at the beginning of the bend. The origin of the coordinates system is on the lower wall at the bend exit proper. The centreline Mach number at the computational inlet is $\breve{M}_{\text{CL}_i} \approxeq 0.097$. The boundary-layers on the duct walls grow downstream, accelerating the flow to $\breve{M}_{\text{CL}} \approxeq 0.1$ just upstream of the bend.

The very thin inlet boundary-layers develop from the computational inflow (x $=$ -1.4365 m; Fig. 4) to the first measurement station (s $=$ -0.185 m; x $=$ -0.166124 m) upstream of the bend (Fig. 4). The profile of the streamwise (wall-parallel) velocity $\bar{u}_s$, as a function of the distance from the wall y_s, is quite well predicted at this station (Fig. 4) by all three turbulence closures, the GLVY RSM performing slightly better than the others. Nonetheless, all models predict a more filled streamwise velocity profile $\bar{u}_s$ near the wall ($y_s \lessapprox 0.005$ m; Fig. 4). This better prediction of the streamwise velocity profile $\bar{u}_s$ by the GLVY RSM is consistent with the improved prediction of the shear Reynolds-stress $\overline{u_s'v_s'}$ in the outer part of the boundary-layer ($y_s \gtrapprox 0.01$ m), all turbulence closures predicting very well $\overline{u_s'v_s'}$ near the wall (Fig. 4). Regarding the streamwise $\overline{u_s'^2}$ and wall-normal $\overline{v_s'^2}$ diagonal Reynolds-stresses, expectedly, the linear LS k–ε model fails, because of the pathological shortcomings of the Boussinesq hypothesis [63, pp.

273–278], predicting quasi-isotropic profiles with a maximum value of ~ 4 near the wall, underpredicting (overpredicting) $\overline{u_s'^2}$ ($\overline{v_s'^2}$) by more than twofold (Fig. 4). The two RSMs, GLVY and WNF–LSS, yield similar results, in good agreement with measurements, but both slightly underpredict the diagonal Reynolds-stresses ($\overline{u_s'^2}$ and $\overline{v_s'^2}$) ($y_s \lessapprox 0.005$ m).

The experimental investigation [53] focusses on the response and subsequent relaxation of the boundary-layer, on the convex side of the bend (lower wall), after the strong perturbation by the extra strain induced by the short region of curvature (on the convex side of the bend the radius of curvature is $r_c \approxeq 6\delta_0$, where $\delta_0 \approxeq 22$ mm is the boundary-layer thickness upstream of the bend inlet [53, Tab. 1, p. 214], and the curvilinear length of the bend is $\sim 3\delta_0$). At the first measurement station downstream of the bend exit ($x = 0.030$ m $\approxeq 1.4\delta_0$; Fig. 5) all of the models yield very similar results for the streamwise mean velocity $\bar{u}$ profile (for $x \geq 0$, downstream of the bend exit located at $x = 0$, x is the streamwise wall-parallel direction), the computed profiles being more filled than the experimental data, possibly a consequence of the slight discrepancy in the incoming velocity profile ($x = -0.166124$ m, $y_s \lessapprox 0.005$ m; Fig. 4). On the other hand, there are substantial differences in the prediction of the Reynolds stresses ($x = 0.030$ m $\approxeq 1.4\delta_0$; Fig. 5), between the two RSMs, on the one hand, and the linear LS k–ε on the other. Both RSMs predict very accurately the shear Reynolds-stress profile at the bend exit ($x = 0.030$ m $\approxeq 1.4\delta_0$; Fig. 5) whereas the linear LS k–ε largely fails to predict the strong reduction of $\overline{u'v'}$ in the outer part of the boundary layer ($y \gtrapprox 0.4\delta$) induced by the convex curvature effect [53, pp. 211–212]. The predictions of the Reynolds-stresses by the two RSMs at the bend exit ($x = 0.030$ m $\approxeq 1.4\delta_0$; Fig. 5) are in good agreement with one another and with the experimental data; in the region close to the wall where the two RSMs differ no experimental data were available. Further downstream, in the relaxation and recovery regions ($x \in \{0.183\,\text{m}, 0.335\,\text{m}, 1.250\,\text{m}\}$; Fig. 5), the GLVY RSM predicts quite accurately the streamwise mean velocity $\bar{u}$ profiles, whereas the WNF LSS RSM yields results very similar to the linear LS k–ε, both underpredicting $\bar{u}$ near the wall in the recovery region ($y \lessapprox 0.005$ m, $x \in \{0.183\,\text{m}, 0.335\,\text{m}\}$; Fig. 5). Regarding the prediction of the Reynolds-stresses, both RSMs are globally in good agreement with one another and with measurements. In the relaxation region the WNF LSS RSM is in slightly better agreement with measurements in the outer part of the boundary-layer ($0.005 \lessapprox y \lessapprox 0.01$ m, $x \in \{0.183\,\text{m}, 0.335\,\text{m}\}$; Fig. 5), whereas the GLVY RSM is slightly better in the recovery region ($x = 1.250$ m; Fig. 5), which is quite challenging to predict [53, pp. 225–230]. The prediction of the Reynolds-stresses everywhere (Fig. 5) highlights the inaccuracy of the linear LS k–ε model, compared to RSMs, in simulating this test-case.

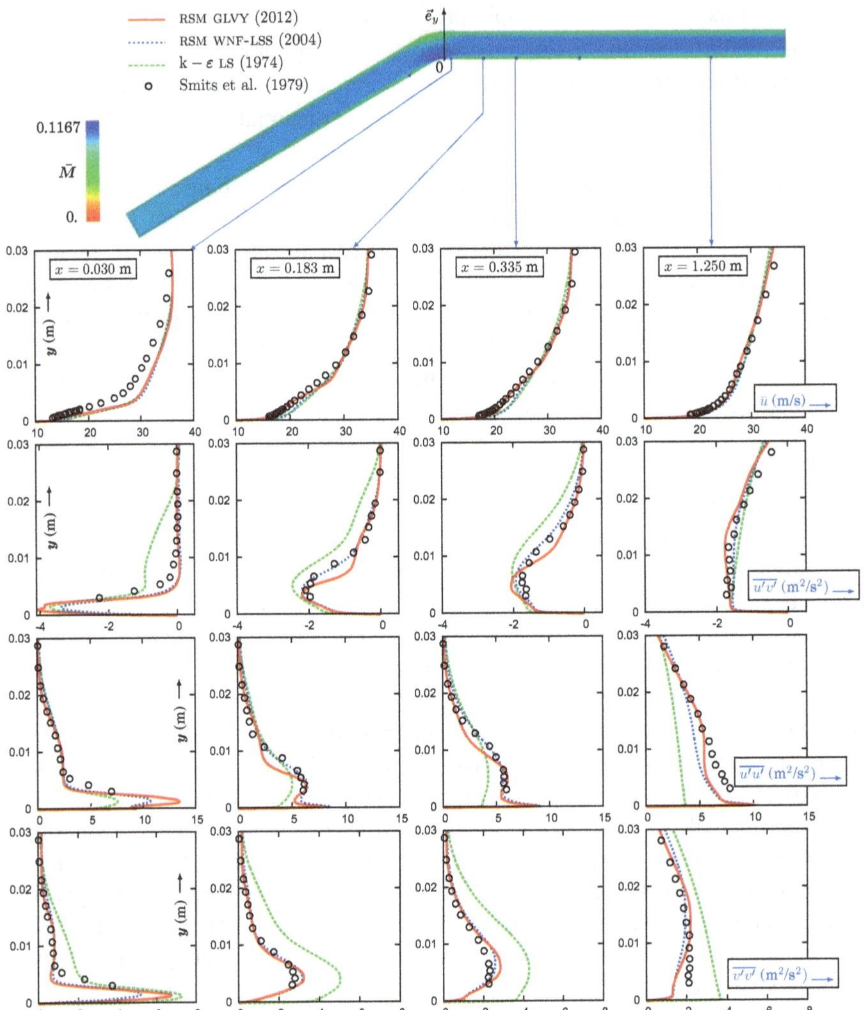

Fig. 5 Comparison of measured profiles of streamwise velocity $\tilde{u}$ and Reynolds-stresses ($\overline{u'^2}$, $\overline{v'^2}$, $\overline{u'v'}$), on the convex side (*lower wall*) of the Smits et al. [53] 30 deg bend (downstream of the bend exit), with computations using the GLVY [26] and WNF–LSS [20] RSMs and the linear LS k–ε [35] model ($1,025 \times 385$ grid [45]; level plots of Mach number $\breve{M}$ using the GLVY RSM)

All of the models predict quite well the pressure coefficient[3] C_p along the lower wall of the duct (Fig. 6). On the lower wall (convex side) the WNF–LSS RSM is

[3]The measured C_p is defined with respect to a reference velocity measured slightly inside the contraction [53, Fig 1, p. 211] of $31.9\,\text{m}\,\text{s}^{-1}$ [45] and to a reference pressure at the same location [53, p. 213], so that at the duct inlet (computational inflow) the measured $C_p \approxeq -0.07$ [53, p. 213]. Therefore, following [45] C_p computed with respect to the static pressure at the computational inlet was corrected by subtracting 0.07 to compare with measurements (Fig. 6).

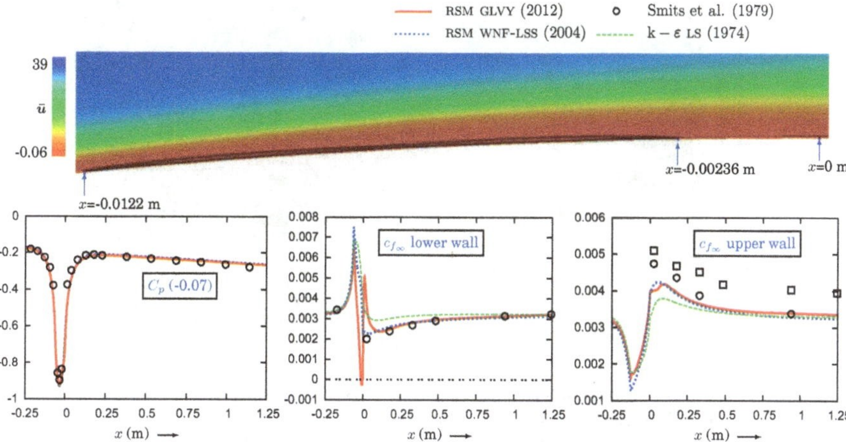

Fig. 6 Comparison of measured x-wise evolution of pressure coefficient C_p on the convex side (lower wall) and skin-friction coefficient c_{f_∞} on the both walls of the Smits et al. [53] 30 deg bend (downstream of the bend exit), with computations using the GLVY [26] and WNF–LSS [20] RSMs and the linear LS k–ε [35] model (1, 025 × 385 grid [45]; level plots of x-wise velocity $\bar{u}$ using the GLVY RSM)

in very good agreement with experimental data for the skin-friction coefficient c_{f_∞} (Fig. 6), contrary to the linear LS k–ε model. The GLVY RSM yields results similar to the WNF–LSS, except for a tiny region of separated flow very near the wall ($-12.2\,\text{mm} \lessapprox x \lessapprox -2.36\,\text{mm}$; Fig. 6), followed by an overshoot at reattachment (probably related to the closure for pressure-diffusion). On the upper wall (concave side), the experiment reveals the presence of streamwise vortices inducing a spanwise varying flow [53], which cannot be predicted by a 2-D RANS computation. The two RSMs are in good agreement one with another and perform better than the linear LS k–ε model (Fig. 6), predicting values close to the lower experimental curve [53, trough, Fig. 4, p. 215].

3.3 NACA 4412 Airfoil Trailing-Edge Separation

This test-case [45] concerns the flow around a 2-D NACA 4412 airfoil at AoA = 13.87 deg (near maximum lift), studied experimentally by Coles and Wadcock [4, 60]. A small separation zone was observed [4, 60] on the upper (suction) side, near the trailing-edge (Fig. 7).

The geometric ($\chi = 0.9012\,\text{m}$) and freestream ($p_{t_\infty} = 98{,}300\,\text{Pa}$, $T_{t_\infty} = 298\,\text{K}$) parameters correspond to the measurements of Coles and Wadcock [4] for $M_\infty = 0.085$ and $Re_\chi = 1.66 \times 10^6$ (Fig. 7). Freestream turbulence intensity at the inflow boundary ($x \sim -175\chi$) was set to $T_{u_\infty} := (\frac{2}{3}\text{k}_\infty)^{\frac{1}{2}} V_\infty^{-1} = \frac{1}{2}\,\%$ with a length scale $\ell_{\text{T}_\infty} := \text{k}_\infty^{\frac{3}{2}}\varepsilon_\infty^{-1} = 0.1\,\text{m} \approxeq 0.11\chi$, resulting [22] to a turbulence

Fig. 7 Level plots of Mach number $\breve{M}$ around the NACA 4412 airfoil [4, 60], computed using the GLVY RSM (481 × 253 O-grid [45]; $M_\infty = 0.085$; AOA = 13.87 deg; $Re_\chi = 1.66 \times 10^6$; transition-trips @2.5 %χ (suction) and @10.3 %χ (pressure) [4]; farfield boundary @175χ; $T_{u_{\text{LE}}} \approxeq 0.05$ %)

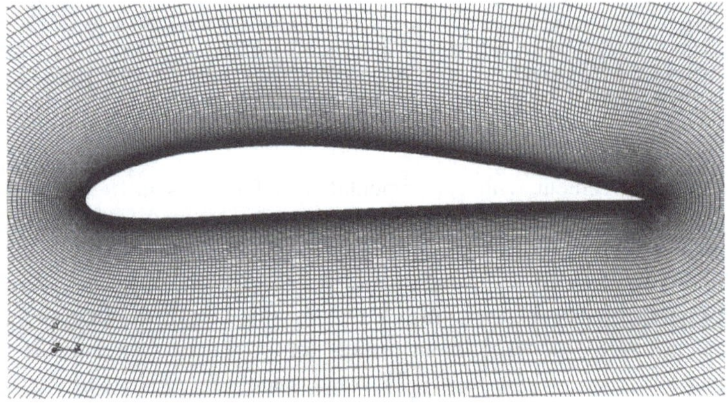

Fig. 8 View of the 481 × 253 points O-grid [45] near the NACA 4412 airfoil [4, 60]

intensity at the leading-edge of $T_{u_{\text{LE}}} \approxeq 0.05$ %. In the experiment [4], the flow was tripped at 2.5 %χ on the upper (suction) side and at 10.3 %χ on the lower (pressure) side. In the computations we used trip zones spanning 2.5 %$\chi \pm 5$ mm on the upper (suction) side and 10.3 %$\chi \pm 5$ mm on the lower (pressure) side, with trip-region height $\delta_{\text{TRIP}} = 1$ mm. The tripping methodology of Carlson [2] and Pandya et al. [42] extended to a second-moment-closure framework [15] was used, injecting, when appropriate, turbulence with local intensity $T_{u_{\text{TRIP}}} = 0.30$. The computations were run on a 481 × 253 O-grid (481 points on the airfoil surface; Fig. 8) which has equivalent resolution near the airfoil as the 897 × 257 grid of the NASA Turbmodels website [45]. For the flow conditions of the test-case the wall-normal size of the first grid-cell adjacent to the wall is $\Delta n_w^+ \approxeq 0.1$ Computed (GLVY RSM) standard lift and drag coefficients based on the freestream conditions were $c_L = 1.621$

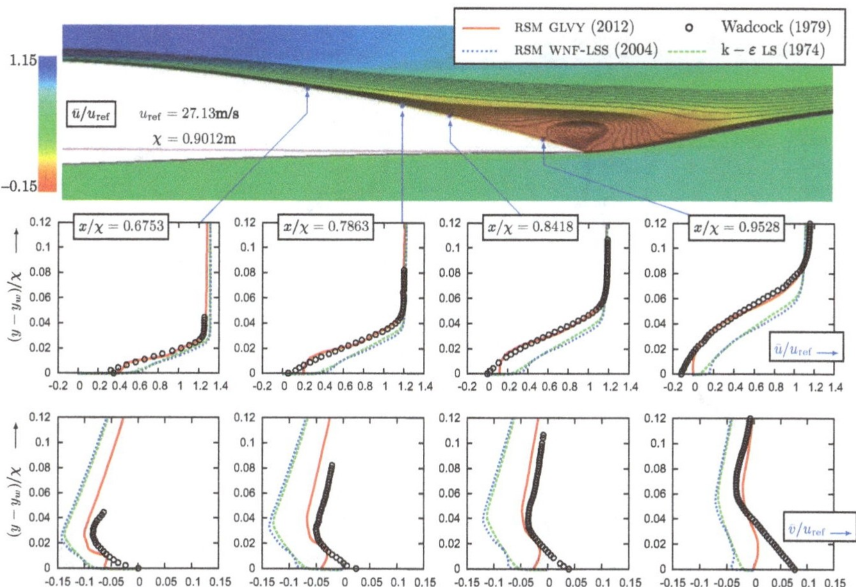

Fig. 9 Comparison of measured [60] profiles of mean-flow velocity components ($\bar{u}$ and $\bar{v}$; chord-aligned coordinates) as a function of the chord-normal distance from the upper (suction) surface of the NACA 4412 airfoil [4], at 4 measurement stations near the trailing-edge, with computations using the GLVY [26] and WNF–LSS [20] RSMs and the linear LS k–ε [35] model (481 × 253 O-grid [45]; $M_\infty = 0.085$; AOA $= 13.87$ deg; $Re_\chi = 1.66 \times 10^6$; transition-trips @2.5 %χ (suction) and @10.3 %χ (pressure) [4]; farfield boundary @175χ; $T_{u_{LE}} \approxeq 0.05$ %; level plots of $\bar{u}$ using the GLVY RSM)

and $c_D = 0.0312$. The computations converged reasonably well to a steady state (residual c_D fluctuation of ±0.5 counts).

The WNF–LSS RSM and the linear LS k–ε model yield quite similar predictions for the mean-velocity profiles near the trailing-edge of the airfoil (Fig. 9), which are not in very good agreement with measurements. On the contrary, the GLVY RSM, which has an optimized rapid redistribution closure $\phi_{ij}^{(\text{RH})}$ (3i) and a model for the pressure-diffusion term d_{ij}^{p} (3c) is in better agreement with experimental data (Fig. 9), especially in the outer part of the boundary-layer, although separation is predicted $\sim 2\,\%\chi$ downstream of the experimental location. As a consequence, the maximum backflow velocity is $\sim -0.15V_{\text{ref}}$ instead of $-0.2V_{\text{ref}}$. In line with the predictions of mean-flow velocities (Fig. 9), the WNF–LSS RSM and the linear LS k–ε model yield quite similar predictions for the shear Reynolds-stress $\overline{u'v'}$ (Fig. 10), while the WNF–LSS RSM predicts slightly better the diagonal stresses ($\overline{u'^2}$ and $\overline{v'^2}$; Fig. 10). The GLVY RSM yields the best agreement with experimental data for the Reynolds-stresses ($\overline{u'^2}$, $\overline{v'^2}$ and $\overline{u'v'}$; Fig. 10), predicting the correct y-wise location of the maximum peak for all the components, at each measurement station. However, the levels of $\overline{u'^2}$ and $\overline{u'v'}$ are underestimated near the trailing edge

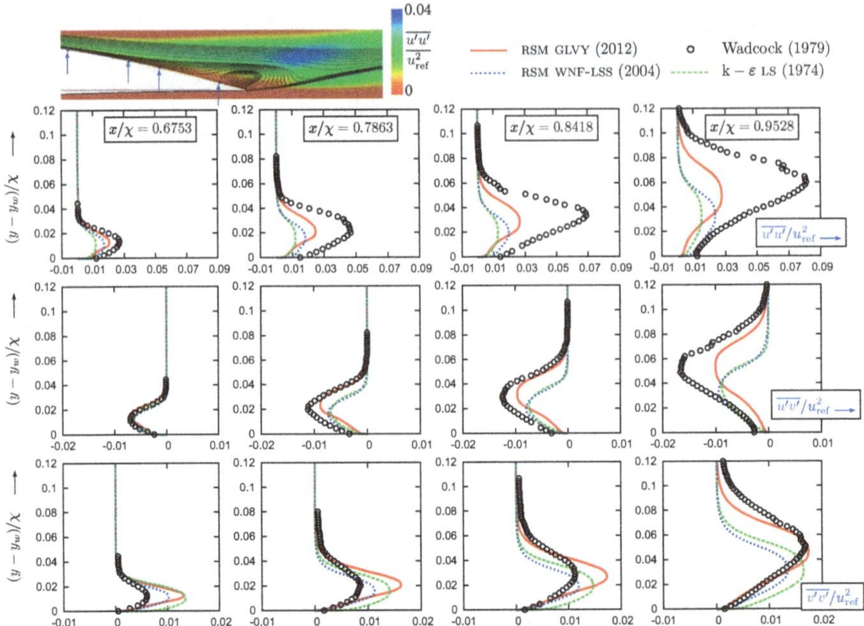

Fig. 10 Comparison of measured [60] profiles of Reynolds-stresses ($\overline{u'^2}$, $\overline{v'^2}$ and $\overline{u'v'}$; chord-aligned coordinates) as a function of the chord-normal distance from the upper (suction) surface of the NACA 4412 airfoil [4], at four measurement stations near the trailing-edge, with computations using the GLVY [26] and WNF–LSS [20] RSMs and the linear LS k–ε [35] model (481 × 253 O-grid [45]; $M_\infty = 0.085$; AOA = 13.87 deg; $Re_\chi = 1.66 \times 10^6$; transition-trips @2.5 %χ (suction) and @10.3 %χ (pressure) [4]; farfield boundary @175χ; $T_{u_{\text{LE}}} \approxeq 0.05\,\%$; level plots of $\overline{u'^2}$ using the GLVY RSM)

($x = 0.7863\chi$; Fig. 10), while $\overline{v'^2}$ is overestimated at the beginning of separation ($x \in \{0.7863\chi, 0.8418\chi\}$; Fig. 10).

3.4 3-D Supersonic Square Duct

Experimental data for the 3-D developing turbulent supersonic flow in a square duct test-case [45] were obtained by Davis and Gessner [5]. This is a straight duct of square cross-section (width $D = 25.4$ mm), 1.27 m $= 50D$ long (the computational domain extends slightly downstream to $52D$). Computations were run on the two grids from [45], of 3×10^6 ($481 \times 81 \times 81$) and 25×10^6 ($961 \times 161 \times 161$) points (Fig. 11), which discretize $\frac{1}{4}$ of the duct with y-wise and z-wise symmetry conditions. At inflow, uniform supersonic Mach number $M_i = 3.9$, total pressure $p_{t_i} = 416{,}000$ Pa and total temperature $T_{t_i} = 300.15$ K, with turbulence intensity $T_{u_i} = 1\,\%$ and length scale $\ell_{T_i} = 50$ mm (which is roughly twice the duct's height)

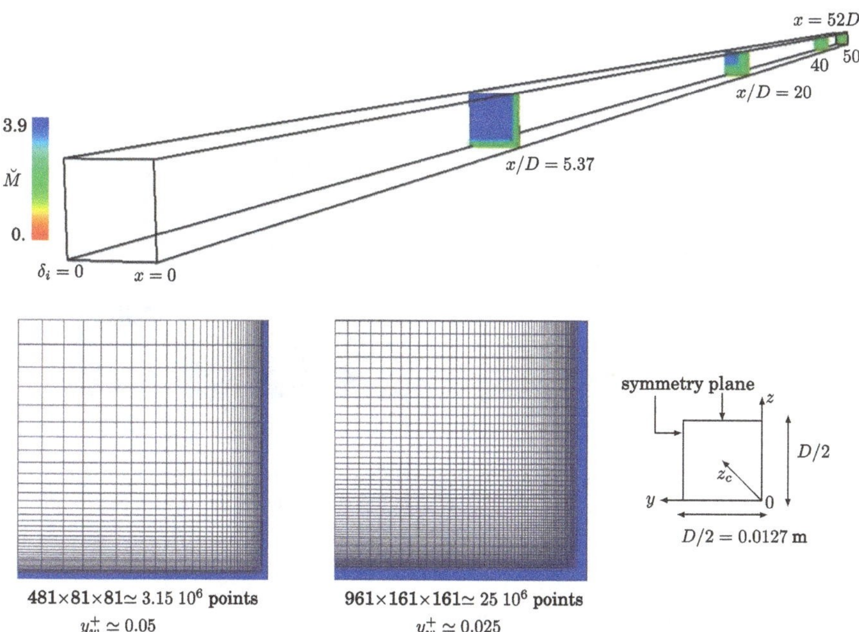

Fig. 11 Geometry and computational grids ($481 \times 81 \times 81$ of 3×10^6 points and $961 \times 161 \times 161$ of 25×10^6 points [45]) discretizing $\frac{1}{4}$ of the Davis and Gessner [5] supersonic square duct (levels of Mach number $\breve{M}$ computed on the 25×10^6 points grid using the GLVY RSM).

were applied.[4] The initial inflow boundary-layer thickness was $\delta_i = 0\,\text{mm}$, and adiabatic no-slip conditions were applied at the walls. For these conditions, the wall-normal size of the first grid-cell adjacent to the wall is $\Delta y_w^+ \approxeq \Delta z_w^+ \approxeq 0.05$ for the coarser 3×10^6 points grid and $\Delta y_w^+ \approxeq \Delta z_w^+ \approxeq 0.025$ for the finer 25×10^6 points grid. As the boundary-layers on the duct walls grow downstream, they induce blockage which decelerates the supersonic flow to ~2 at the duct's exit ($x = 50D$).

The main challenge of this test-case is to predict the secondary flows induced mainly by turbulence anisotropy, which can only be predicted by anisotropy-resolving closures. At the duct exit ($x = 50D$) the GLVY RSM predicts two counter-rotating vortices with a strong inflow toward the corner along the corner-bisector traverse which is then evacuated along the two walls, in excellent agreement with experimental data (Fig. 12). Comparison of the GLVY RSM results on the two grids ($481 \times 81 \times 81$ and $961 \times 161 \times 161$ [45]; Fig. 12) indicates that results on the coarser $481 \times 81 \times 81$ grid are reasonably grid-converged (Figs. 13–15).

The GLVY RSM is globally in good agreement with experimental data for the streamwise velocity $\tilde{u}$ (Fig. 13), contrary to the linear LS k–ε model that overpredicts

[4]For these flow conditions, in agreement with measurements, the unit-Reynolds-number at inflow is $Re_{1_i} \approxeq 19.66 \times 10^6$.

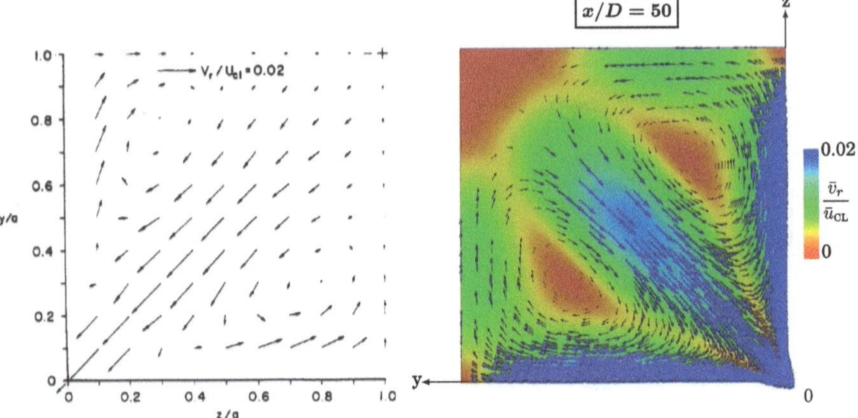

Fig. 12 Comparison of measured [5] secondary velocity $\tilde{v}_r$ vectors, nondimensionalized by the local centerline velocity $\tilde{u}_{CL}(x)$, at $x = 50D$, with computations using the GLVY RSM [26] (961 × 161 × 161 grids [45]; Fig. 12)

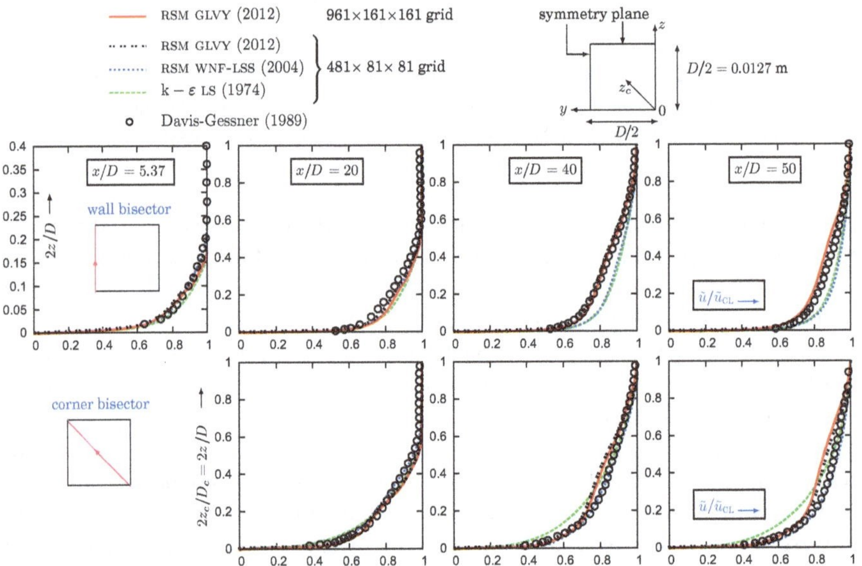

Fig. 13 Comparison of measured [5] streamwise velocity $\tilde{u}$ profiles, nondimensionalized by the local centreline velocity $\tilde{u}_{CL}(x)$, along the wall-bisector z ($y = \frac{1}{2}D$) and along the corner-bisector z_c ($y = z$), at 4 measurement stations, with computations using the GLVY [26] and WNF–LSS [20] RSMs and the linear LS k–ε [35] model (481 × 81 × 81 and 961 × 161 × 161 grids [45]; Fig. 12)

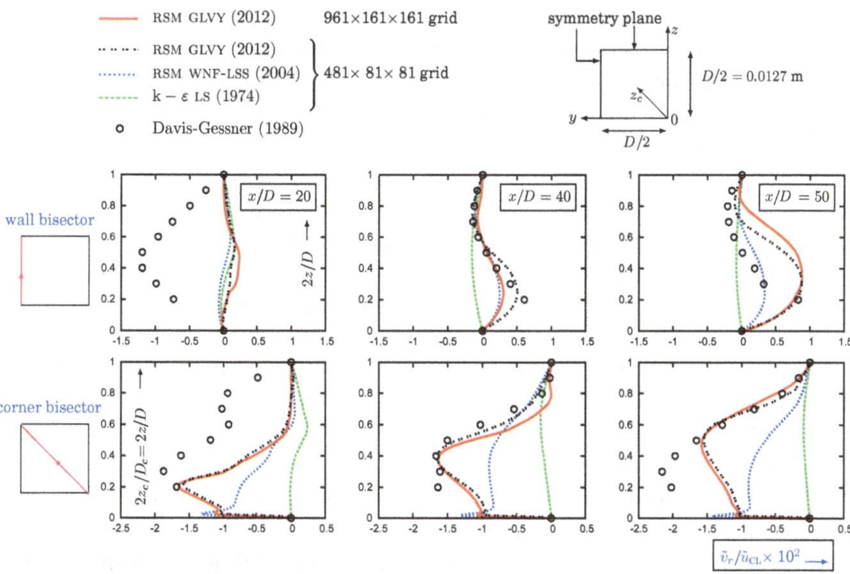

Fig. 14 Comparison of measured [5] secondary velocity $\tilde{v}_r$ profiles, nondimensionalized by the local centreline velocity $\tilde{u}_{\text{CL}}(x)$, along the wall-bisector z $(y = \frac{1}{2}D)$ and along the corner-bisector z_c $(y = z)$, at 4 measurement stations, with computations using the GLVY [26] and WNF–LSS [20] RSMs and the linear LS k–ε [35] model ($481 \times 81 \times 81$ and $961 \times 161 \times 161$ grids [45]; Fig. 12)

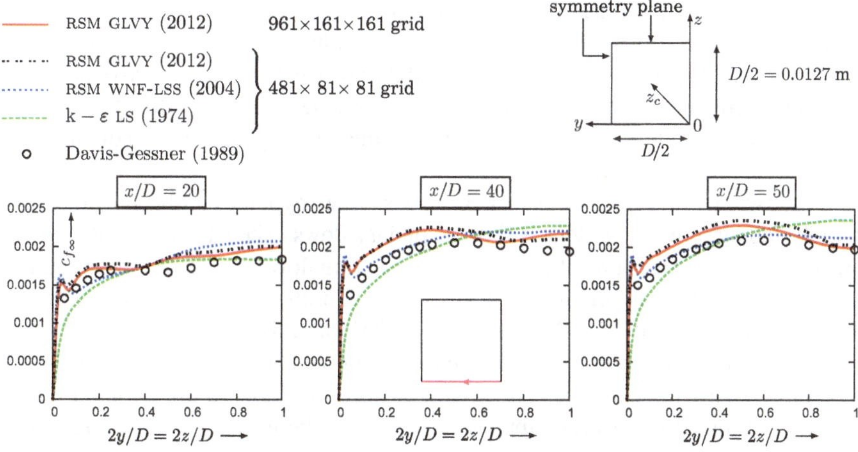

Fig. 15 Comparison of measured [5] skin-friction coefficient c_{f_∞}, along the duct's side, at 3 measurement stations, with computations using the GLVY [26] and WNF–LSS [20] RSMs and the linear LS k–ε [35] model ($481 \times 81 \times 81$ and $961 \times 161 \times 161$ grids [45]; Fig. 12)

$\tilde{u}$ along the wall-bisector and severely underpredicts it along the corner-bisector (Fig. 13). The WNF–LSS RSM is slightly better than the GLVY RSM along the corner-bisector (away from the corner; Fig. 13), but overpredicts $\tilde{u}$ along the wall-bisector where it is very close to the LS k–ε results (Fig. 13).

The prediction of secondary (in-plane $\perp x$) velocities (Fig. 12) is much more challenging. The linear LS k–ε [35] model, because of the pathological shortcomings of the Boussinesq hypothesis [63, pp. 273–278], completely fails (Fig. 14) predicting very weak secondary velocities, contrary to the RSMs. Overall, the GLVY RSM is in better agreement with measurements (some discrepancies near the wall notwithstanding; Fig. 14), except in the outer part of the boundary-layer along the wall-bisector at $x = 50D$ (Fig. 14). Along the corner-bisector the GLVY RSM is in better agreement with measurements than the WNF–LSS RSM, which underpredicts the secondary flow velocities (Fig. 14).

Regarding skin-friction (Fig. 15) the two RSMs are in very good agreement with experimental data, contrary to the linear LS k–ε, the more so with increasing x. The experimental distributions of c_{f_∞} invariably present a maximum followed by an inflection point, that move away from the corner with increasing x (Fig. 15). This particular feature is well predicted by the GLVY RSM (Fig. 15), despite a slight overestimation of the c_{f_∞}.

4 Conclusions

The GLVY RSM [26] was assessed against experimental measurements for four test-cases from NASA Turbmodels website [45]. Results were also presented for the linear LS k–ε [35] (to put into perspective the improvements in predictive accuracy by using differential RSMs) and with the baseline WNF–LSS RSM [20] (to highlight improvements by more elaborate closures of the velocity/pressure-gradient tensor Π_{ij}).

Because of the shortcomings of the Boussinesq hypothesis, the linear LS k–ε largely fails in predicting detached or secondary flows. The baseline WNF–LSS RSM substantially improves upon the LS k–ε in predicting flows dominated by streamline curvature or secondary motions but yields results rather similar to the k–ε model for separated flows. It is well established [9, 26] that this unsatisfactory behaviour of the WNF–LSS RSM is related to the closure adopted for the pressure terms Π_{ij}. The GLVY RSM which includes an explicit closure for pressure-diffusion term $d_{ij}^{(p)}$ and an optimized rapid redistribution model gives better overall agreement with measurements.

Nonetheless, there is room for improvement because the GLVY RSM slightly underpredicts trailing-edge separation and overpredicts curvature effects (c_{f_∞} overshoot on convex bend). Both these issues seem to be related to the near-wall behaviour of the model, and require better near-wall modelling of the velocity/pressure-gradient tensor Π_{ij}. An interesting possibility is the development of a full differential r_{ij}–ε_{ij} closure which is the subject of on-going work.

The inclusion of ε_{ij} as additional variables seems the only possibility for the satisfactory a posteriori prediction of the near-wall turbulence anisotropy. The availability of the ε_{ij} independently of r_{ij} (contrary to algebraic models for ε_{ij}) offers an extended tensorial representation basis for Π_{ij} which, it is hoped, will allow less stringent near-wall dampings in the model.

Acknowledgements The computations were performed using HPC resources from GENCI–IDRIS (Grant 2014-020218) and from ICS–UPMC (ANR–10–EQPX–29–01). The authors are listed alphabetically.

References

1. Ben Nasr N, Gerolymos GA, Vallet I (2014) Low-diffusion approximate Riemann solvers for Reynolds-stress transport. J Comput Phys 268:186–235. doi:10.1016/j.jcp.2014.02.010
2. Carlson JR (1997) Applications of algebraic Reynolds-stress turbulence models—part 1: incompressible flat plate. J Propuls Power 13:610–619
3. Chou PY (1945) On velocity correlations and the solutions of the equations of turbulent fluctuations. Q Appl Math 3:38–54
4. Coles D, Wadcock AJ (1979) Flying-hot-wire study of flow past a naca 4412 airfoil at maximum lift. AIAA J 17(4):321–329
5. Davis DO, Gessner FB (1989) Further experiments on supersonic turbulent flow development in a square duct. AIAA J 27(8):1023–1030
6. Gerolymos GA (1990) Implicit multiple-grid solution of the compressible Navier-Stokes equations using $\text{k}-\varepsilon$ turbulence closure. AIAA J 28(10):1707–1717
7. Gerolymos GA, Vallet I (1996) Implicit computation of the 3-D compressible Navier-Stokes equations using $\text{k}-\varepsilon$ turbulence closure. AIAA J 34(7):1321–1330
8. Gerolymos GA, Vallet I (1997) Near-wall Reynolds-stress 3-D transonic flows computation. AIAA J 35(2):228–236
9. Gerolymos GA, Vallet I (2001) Wall-normal-free near-wall Reynolds-stress closure for 3-D compressible separated flows. AIAA J 39(10):1833–1842
10. Gerolymos GA, Vallet I (2002) Wall-normal-free Reynolds-stress model for rotating flows applied to turbomachinery. AIAA J 40(2):199–208
11. Gerolymos GA, Vallet I (2005) Mean-flow-multigrid for implicit Reynolds-stress-model computations. AIAA J 43(9):1887–1898
12. Gerolymos GA, Vallet I (2007) Robust implicit multigrid Reynolds-stress-model computation of 3-D turbomachinery flows. ASME J Fluids Eng 129(9):1212–1227
13. Gerolymos GA, Vallet I (2009) aerodynamics (a library and software package for computational aerodynamics). http://sourceforge.net/projects/aerodynamics
14. Gerolymos GA, Vallet I (2009) Implicit mean-flow-multigrid algorithms for Reynolds-stress-model computations of 3-D anisotropy-driven and compressible flows. Int J Numer Methods Fluids 61(2):185–219. doi:10.1002/fld.1945
15. Gerolymos GA, Vallet I (2013) Bypass transition and tripping in Reynolds-stress model computations. In: AIAA Paper 2013-2425, 21st AIA Computational Fluid Dynamics Conference, San Diego, 24–27 June 2013
16. Gerolymos GA, Vallet I (2014) Pressure, density, temperature and entropy fluctuations in compressible turbulent plane channel flow. J Fluid Mech 757:701–746. doi:10.1017/jfm.2014.431
17. Gerolymos GA, Kallas YN, Papailiou KD (1989) The behaviour of the normal fluctuation terms in the case of attached and detached turbulent boundary-layers. Revue de Physique Appliquée (Paris) 24(3):375–387

18. Gerolymos GA, Michon GJ, Neubauer J (2002) Analysis and application of chorochronic periodicity for turbomachinery rotor/stator interaction computations. J Propuls Power 18(2):1139–1152
19. Gerolymos GA, Neubauer J, Sharma VC, Vallet I (2002) Improved prediction of turbomachinery flows using near-wall Reynolds-stress model. ASME J Turbomach 124(1):86–99
20. Gerolymos GA, Sauret E, Vallet I (2004) Contribution to the single-point-closure Reynolds-stress modelling of inhomogeneous flow. Theor Comput Fluid Dyn 17(5–6):407–431
21. Gerolymos GA, Sauret E, Vallet I (2004) Oblique-shock-wave/boundary-layer interaction using near-wall Reynolds-stress models. AIAA J 42(6):1089–1100
22. Gerolymos GA, Sauret E, Vallet I (2004) Influence of inflow-turbulence in shock-wave/turbulent-boundary-layer interaction computations. AIAA J 42(6):1101–1106
23. Gerolymos GA, Sénéchal D, Vallet I (2009) Very-high-order WENO schemes. J Comput Phys 228:8481–8524. doi:10.1016/j.jcp.2009.07.039
24. Gerolymos GA, Joly S, Mallet M, Vallet I (2010) Reynolds-stress model flow prediction in aircraft-engine intake double-S-shaped duct. J Aircr 47(4):1368–1381. doi:10.2514/1.47538
25. Gerolymos GA, Lo C, Vallet I (2012) Tensorial representations of Reynolds-stress pressure-strain redistribution. ASME J Appl Mech 79(4):044506. doi:10.1115/1.4005558
26. Gerolymos GA, Lo C, Vallet I, Younis BA (2012) Term-by-term analysis of near-wall second moment closures. AIAA J 50(12):2848–2864. doi:10.2514/1.J051654
27. Gerolymos GA, Sénéchal D, Vallet I (2013) Wall effects on pressure fluctuations in turbulent channel flow. J Fluid Mech 720:15–65. doi:10.1017/jfm.2012.633
28. Gessner FB, Emery AF (1981) The numerical prediction of developing turbulent flow in rectangular ducts. ASME J Fluids Eng 103:445–455. doi:10.1080/10618562.2013.772984
29. Gibson MM, Launder BE (1978) Ground effects on pressure fluctuations in the atmospheric boundary-layer. J Fluid Mech 86:491–511
30. Hanjalić K (1994) Advanced turbulence closure models: A view of current status and future prospects. Int J Heat Fluid Flow 15:178–203
31. Hanjalić K, Launder BE (1976) Contribution towards a Reynolds-stress closure for low-Reynolds-number turbulence. J Fluid Mech 74:593–610
32. Jakirlić S, Eisfeld B, Jester-Zürker R, Kroll N (2007) Near-wall Reynolds-stress model calculations of transonic flow configurations relevant to aircraft aerodynamics. International Journal of Heat and Fluid Flow 28:602–615.
33. Klebanoff PS (1955) Characteristics of turbulence in a boundary-layer with zero pressure gradient. Report 1247, NACA
34. Launder BE, Reece GJ, Rodi W (1975) Progress in the development of a Reynolds-stress turbulence closure. J Fluid Mech 68:537–566
35. Launder BE, Sharma BI (1974) Application of the energy dissipation model of turbulence to the calculation of flows near a spinning disk. Lett Heat Mass Transf 1:131–138
36. Launder BE, Shima N (1989) 2-moment closure for the near-wall sublayer: Development and application. AIAA J 27(10):1319–1325
37. Lumley JL (1978) Computational modeling of turbulent flows. Adv Appl Mech 18:123–176
38. Manceau R, Wang M, Laurence D (2001) Inhomogeneity and anisotropy effects on the redistribution term in Reynolds-averaged Navier-Stokes modelling. J Fluid Mech 438:307–338
39. Naot D, Shavit A, Wolfshtein M (1970) Interactions between components of the turbulent velocity correlation tensor due to pressure fluctuations. Israel J Technol 8(3):259–269
40. Naughton JW, Sheplak M (2002) Modern developments in shear-stress measurement. Prog Aerosp Sci 38:515–570

41. Österlund JM (1999) Experimental studies of zero-pressure-gradient turbulent boundary-layer flow. Doctoral thesis, Royal Institute of Technology (KTH), Stockholm, Sweden. TRITA-MEK Tech. Rep. 1999:16, ISSN 0348–467X ISRN KTH/MEK/TR–99/16–SE
42. Pandya M, Abdol-Hamid K, Campbell R, Frink N (2006) Implementation of flow tripping capability in the USM3D unstructured flow solver. In: AIAA Paper 2006-0919, 44th AIAA Aerospace Sciences Meeting and Exhibit, Reno, 9–12 January 2006
43. Rotta J (1951) Statistische Theorie nichthomogener Turbulenz — 1. Mitteilung. Zeitschrift für Physik 129:547–572
44. Rumsey CL (2007) Apparent transition behavior of widely-used turbulence models. Int J Heat Fluid Flow 28:1460–1471
45. Rumsey CL (2010) NASA Langley Research Center Turbulence Modeling Resource. http://turbmodels.larc.nasa.gov/index.html, visited November 2014
46. Rumsey CL, Smith BR, Huang GP (2010) Description of a website resource for turbulence model verification and validation. In: AIAA Paper 2010-4742, 40th AIAA Fluid Dynamics Conference, Chicago, 28 June–1 July 2010
47. Sauret E, Vallet I (2007) Near-wall turbulent pressure diffusion modelling and influence in 3-D secondary flows. ASME J Fluids Eng 129(5):634–642
48. Schlatter P, Li Q, Brethouwer G, Johansson AV, Henningson DS (2010) Simulations of spatially evolving turbulent boundary-layers up to $re_\theta = 4300$. Int J Heat Fluid Flow 31:251–261. doi:10.1016/j.ijheatfluidflow.2009.12.011
49. Shir CC (1973) A preliminary numerical study of atmospheric turbulent flows in the idealized planetary boundary-layer. J Atmos Sci 30:1327–1339
50. Sillero JA, Jiménez J, Moser RD (2013) One-point statistics for turbulent wall-bounded flows at Reynolds numbers up to $\delta^+ \approx 2000$. Phys Fluids 25:105102. doi:10.1063/1.4823831
51. Simpson RL (1989) Turbulent boundary-layer separation. Annu Rev Fluid Mech 21:205–234
52. Smith RW (1994) Effect of Reynolds number on the structure of turbulent boundary-layers. PhD thesis, Princeton University, Princeton
53. Smits AJ, Young STB, Bradshaw P (1979) The effect of high surface curvature on turbulent boundary-layers. J Fluid Mech 94:209–242
54. So RMC, Zhang HS, Gatski TB, Speziale CG (1994) Logarithmic laws for compressible turbulent boundary-layers. AIAA J 32:2162–2168
55. Speziale CG, Sarkar S, Gatski TB (1991) Modelling the pressure-strain correlation of turbulence: An invariant dynamical systems approach. J Fluid Mech 227:245–272
56. Vallet I (2007) Reynolds-stress modelling of 3-D secondary flows with emphasis on turbulent diffusion closure. ASME J Appl Mech 74(6):1142–1156
57. Vallet I (2008) Reynolds-stress modelling of $M = 2.25$ shock-wave/turbulent-boundary-layer interaction. Int J Numer Methods Fluids 56(5):525–555
58. Vincenti P, Klewicki J, Morrill-Winter C, White CM, Wosnik M (2013) Streamwise velocity statistics in turbulent boundary-layers that spatially develop to high reynolds numbers. Exp Fluids 54:1629. doi:10.1007/s00348-013-1629-9
59. Vos JB, Rizzi A, Darracq D, Hirschel EH (2002) Navier-Stokes solvers in European aircraft design. Prog Aerosp Sci 38:601–697
60. Wadcock AJ (1979) Structure of the turbulent separated flow around a stalled airfoil. In: Contr. Rep. NASA–CR–1979-152263, NASA, Ames Research Center, Moffett Field
61. Wei T, Schmidt R, McMurtry P (2005) Comment on the Clauser-chart method for determining the friction velocity. Exp Fluids 38:695–699
62. Wilcox DC (1988) Reassessment of the scale-determining equation for advanced turbulence models. AIAA J 26:1299–1310
63. Wilcox DC (1998) Turbulence modelling for CFD, 2nd edn. DCW Industries, La Cañada

Modeling of Reynolds-Stress Augmentation in Shear Layers with Strongly Curved Velocity Profiles

René-Daniel Cécora, Rolf Radespiel, and Suad Jakirlić

Abstract An extension and re-calibration of the differential Reynolds-stress model JHh-v2 is performed, followed by a validation for various test cases of aircraft aerodynamics. The additional sink term within the length-scale equation is influential in shear layers where the velocity profile shows large second derivatives. Besides a backward-facing step flow and a zero-pressure-gradient flat plate, which are used for calibration, a round single-stream jet as well as transonic airfoil and bump flows are simulated with the new model for validation purposes. Especially the simulation of the flow over a transonic bump can be improved with the new model version JHh-v3, furthermore the prediction of the potential core length of the turbulent round jet is improved.

1 Introduction

Numerical flow simulation is an important tool in the development process of aircraft industry. Solving of the *Reynolds-Averaged Navier-Stokes* (RANS) equations is regarded as a good compromise between computational effort and accuracy. The quality of RANS simulations highly depends on the modeling of the Reynolds-stress tensor as a measure for the statistical influence of turbulence on the mean flow.

The most commonly used turbulence models in an industrial environment are so-called eddy-viscosity models (e.g., [13, 21]), which link the Reynolds stresses to the shear rate of the mean flow via an eddy viscosity as a proportionality factor. Mostly one or two transport equations are solved to provide the eddy viscosity. Second-moment closure (SMC) models, on the other hand, directly employ transport equations for the Reynolds stresses, offering a higher potential for an adequate

R.-D. Cécora (✉) • R. Radespiel
Institute of Fluid Mechanics, Technische Universität Braunschweig, Hermann-Blenk-Str. 37, 38108 Braunschweig, Germany
e-mail: r-d.cecora@tu-bs.de

S. Jakirlić
Institute of Fluid Mechanics and Aerodynamics, Technische Universität Darmstadt, Alarich-Weiss-Str. 10, 64287 Darmstadt, Germany

B. Eisfeld (ed.), *Differential Reynolds Stress Modeling for Separating Flows in Industrial Aerodynamics*, Springer Tracts in Mechanical Engineering,
DOI 10.1007/978-3-319-15639-2_5

prediction of complex flows. They are often referred to as Reynolds-stress models (RSM).

Recently Cécora et al. [3] presented a SMC model (JHh-v2) which showed promising results in different aeronautical applications, containing high-lift airfoil flow, shock/boundary-layer interaction, and vortex flow. It was however noticed that the JHh-v2 model underestimates the development of turbulence in shear layers with inflection point in the velocity profile. A corresponding example is the backward-facing step (BFS) flow, in which the slowly developing turbulence in the separated shear layer causes a lack of exchange of momentum, leading to an overestimated separation length [16]. Likewise the round-turbulent-jet flow is concerned, in which a reduced spreading of the turbulent shear layer is simulated, exhibiting an overestimated jet core length. Reuß et al. [18] report on the rapid decay of turbulence in the flap cove of a two-element airfoil when simulating with the JHh-v2 model, resulting in instabilities of the cove flow which have an influence on the flap separation mechanisms.

A remedy is found by implementing an additional sink term into the length-scale equation, which is sensitized towards shear flows with inflection point. An appropriate formulation was found within the so-called Scale Adaptive Simulation (SAS) concept [14], which uses an additional source term in the length-scale equation to enhance instabilities by locally reducing turbulence. Contrary to the SAS concept, the additional term is used as a sink term in this work which locally enhances turbulence. The successful application of this new sink term to a similar Reynolds-stress model (RSM-P_{SAS}) has already been shown by Maduta [11] and Jakirlić and Maduta [9]. Both RSM, JHh-v2 and RSM-P_{SAS}, are evolved from the JH model [8], they mainly differ in the length-scale equation, with the JHh-v2 employing the homogeneous dissipation rate ε^h as length-scale variable while the RSM-P_{SAS} model uses the homogeneous part of the inverse time scale $\omega^h = \varepsilon^h/k$. Furthermore the JHh-v2 model contains an additional source term in the length-scale equation which sensitizes the model towards adverse pressure gradients. Last but not least, the models use a different set of coefficients in the length-scale equation.

In this work, the extension and calibration of the JHh-v2 model is presented, using a backward-facing step flow and a zero-pressure-gradient flat plate flow as calibration cases. The resulting model is named as JHh-v3. The performance of the JHh-v3 model is validated considering a turbulent round jet, two cases of the flow around the transonic airfoil RAE 2822 and a transonic axisymmetric bump.

2 Turbulence Modeling

Closure of the RANS equation system is achieved with a differential Reynolds-stress turbulence model. It is based on a near-wall RSM by Jakirlić and Hanjalić [8] (JH RSM), which was implemented into the DLR-TAU Code [20] and extended by two additional source terms by Probst and Radespiel [17] (JHh-v1). A re-calibration for different aeronautical test cases was conducted by Cécora et al. [3], resulting in the model version JHh-v2, which is briefly described in Sect. 2.1. The extension of the JHh-v2 model with an additional sink term is presented in Sect 2.2.

2.1 JHh-v2 Reynolds-Stress Turbulence Model

In general form, the compressible Reynolds-stress transport equation reads:

$$\frac{\partial\left(\overline{\rho}\tilde{R}_{ij}\right)}{\partial t}+\frac{\partial}{\partial x_k}\left(\overline{\rho}\tilde{R}_{ij}\tilde{U}_k\right)=\overline{\rho}P_{ij}+\overline{\rho}\Pi_{ij}-\overline{\rho}\varepsilon_{ij}+\overline{\rho}D^{\nu}_{ij}+\overline{\rho}D^{t}_{ij}+\overline{\rho}M_{ij}\ . \tag{1}$$

Only the production term P_{ij} and the viscous diffusion D^{ν}_{ij} can be determined exactly, whereas the remaining terms on the right-hand side of Eq. (1), which describe pressure-strain correlation, dissipation, turbulent diffusion as well as effects of density fluctuations, require modeling.

For the pressure-strain correlation, a quadratic model formulation with an additional wall-reflection model Π^{w}_{ij} according to Gibson and Launder [6] is employed:

$$\overline{\rho}\Pi_{ij}=\overline{\rho}\Pi_{ij,1}+\overline{\rho}\Pi_{ij,2}+\overline{\rho}\Pi^{w}_{ij} \tag{2}$$

$$\overline{\rho}\Pi_{ij,1}=-\varepsilon^{h}\overline{\rho}\left[C_1\tilde{a}_{ij}+C_1'\left(\tilde{a}_{ik}\tilde{a}_{jk}-\frac{1}{3}\delta_{ij}A_2\right)\right] \tag{3}$$

$$\overline{\rho}\Pi_{ij,2}=-C_2\overline{\rho}\left(P_{ij}-\frac{1}{3}P_{kk}\delta_{ij}\right)\ . \tag{4}$$

The model coefficients, including f_w, C_1^w and C_2^w within the wall-reflection model, contain DNS-calibrated damping functions to account for effects of near-wall turbulence.

With the anisotropic dissipation rate tensor divided into a homogeneous part and a non-homogeneous part $\varepsilon_{ij}=\varepsilon^{h}_{ij}+1/2D^{\nu}_{ij}$, an implicit relation is used for ε^{h}_{ij}:

$$\varepsilon^{h}_{ij}=f_s\tilde{R}_{ij}\frac{\varepsilon^{h}}{\tilde{k}}+(1-f_s)\frac{2}{3}\delta_{ij}\varepsilon^{h}\quad\text{with}\quad f_s=1-\sqrt{A E^2}\ . \tag{5}$$

A scalar length-scale equation is employed to provide the homogeneous dissipation rate ε^h:

$$\begin{aligned}\frac{D\varepsilon^h}{Dt}&=-C_{\varepsilon 1}\frac{\varepsilon^h}{\tilde{k}}\tilde{R}_{ij}\frac{\partial\tilde{U}_i}{\partial x_j}-C_{\varepsilon 2}f_{\varepsilon}\frac{\varepsilon^h\tilde{\varepsilon}^h}{\tilde{k}}+C_{\varepsilon 3}\nu\frac{\tilde{k}}{\varepsilon^h}\tilde{R}_{jk}\frac{\partial^2\tilde{U}_i}{\partial x_j\partial x_l}\frac{\partial^2\tilde{U}_i}{\partial x_k\partial x_l}\\&+\frac{\partial}{\partial x_k}\left[\left(\frac{1}{2}\nu\delta_{kl}+C_{\varepsilon}\frac{\tilde{k}}{\varepsilon^h}\tilde{R}_{kl}\right)\frac{\partial\varepsilon^h}{\partial x_l}\right]+S_l+S_{\varepsilon 4}\end{aligned} \tag{6}$$

with S_l and $S_{\varepsilon 4}$ being two additional source terms for sensitizing the equation to effects of non-equilibrium turbulence [17].

The generalized gradient diffusion model by Daly and Harlow [5] is used for the turbulent diffusion tensor, the mass-fluctuation term is neglected.

2.2 Extension of the JHh-v2 Model for Free Shear Flows

In different flow cases it was noted that the current version of the Reynolds-stress model (JHh-v2) tends to underestimate the growth of Reynolds stresses in free shear layers which contain an inflection point in the velocity distribution. As a remedy, an additional sink term is implemented into the length-scale equation with the intention of locally reducing dissipation and hence supporting the development of turbulence. After implementation, the RSM is re-calibrated and named as JHh-v3.

The background of the implemented sink term can be found within the $k - kL$ model of Rotta [19], which employs a transport equation for the quantity kL, with L being an integral length scale of turbulence:

$$kL = \frac{3}{16} \int_{-\infty}^{\infty} R_{ii}\left(\mathbf{x}, r_y\right) dr_y \,. \tag{7}$$

R_{ij} describes the two-point correlation tensor of the velocity fluctuation u'_i, considering the Einstein summation convention for R_{ii}. In the derivation of the modeled kL equation from an exact transport equation, Rotta notices a second production term that is influential on the performance of his turbulence model. Menter and Egorov [14] proposed their own way of modeling this term, furthermore they transformed it to different length-scale variables in order to make it suitable for modern turbulence models [14, 15]. This second production term contains second derivatives of the velocity tensor, which enables the turbulence model to account for additional length scales within the mean velocity field. Combining it with an eddy-viscosity model gave birth to the Scale Adaptive Simulation concept [14], in which the turbulence model is sensitized for resolving instabilities by locally reducing modeled turbulence. The additional source term especially contributes in shear flows with velocity distributions containing inflection points, which is an indicator for a tendency towards instability. Combined with the length-scale equation of modern eddy-viscosity models, the source term reduces the modeled turbulence in the affected region, causing the simulation to develop unsteadiness. Using these turbulence models with SAS-extension in a time-accurate flow solver (URANS) allows for an improved prediction of turbulence in shear layers with a tendency towards instability. Instead of combining the additional source term to an eddy-viscosity model, Maduta and Jakirlić [12] use the SAS concept in combination with a second-moment closure model.

While the SAS concept is intended to resolve a wide part of the turbulent spectrum in unsteady flow regions, the opposite approach followed in this work is to consider a wider spectrum of instabilities within the turbulence model. Therefore the SAS source term is used as a sink term in the length-scale equation of the Reynolds-stress model:

$$\frac{D\varepsilon^h_{(\mathrm{JHh}-v3)}}{Dt} = \frac{D\varepsilon^h_{(\mathrm{JHh}-v2)}}{Dt} - P_{\mathrm{SAS}} \, . \tag{8}$$

This idea was proposed and successfully applied by Maduta [11] using the JH Reynolds-stress model in combination with a ω^h length-scale equation. After transformation of Maduta's term into an ε^h formulation, the implemented sink term reads:

$$P_{\mathrm{SAS}} = C_{\mathrm{SAS},1} \max\left[P_{\mathrm{SAS},1} - P_{\mathrm{SAS},2}, 0\right] \, , \tag{9}$$

with

$$P_{\mathrm{SAS},1} = 1.755\kappa k S^2 \left(\frac{L}{L_{vk}}\right)^{\frac{1}{2}} \tag{10}$$

and

$$P_{\mathrm{SAS},2} = 3\max\left(C_{\mathrm{SAS},2}\frac{(\nabla\varepsilon^h)^2 k^2 + (\nabla k)^2(\varepsilon^h)^2 - 2k\varepsilon^h\nabla\varepsilon^h\nabla k}{(\varepsilon^h)^2}, (\nabla k)^2\right) \, . \tag{11}$$

The formulation contains the turbulent length scale $L = k^{3/2}/\varepsilon^h$ and the 3D generalization of the classical boundary-layer definition of the von Karman length scale $L_{vk} = \kappa S/\left|\nabla^2 U\right|$. This work describes the combination of the sink term in Eqs. (9)–(11) with the JHh-v2 model.

3 Calibration of the Additional Sink Term

The following investigations have been carried out using the DLR-TAU Code [20], a finite-volume solver which solves the compressible Reynolds-averaged Navier-Stokes equations on hybrid unstructured grids with second order accuracy.

Although the model investigated in this work is derived from the same Reynolds-stress model as the one shown by Maduta [11] and Maduta and Jakirlić [9], minor differences can be found in the formulations and in the model coefficients. Therefore the sink term cannot be just transformed from ω^h to ε^h, it has to be calibrated as well. The additional sink term [Eqs. (9)–(11)] contains two coefficients that need to be calibrated: $C_{\mathrm{SAS},1}$ and $C_{\mathrm{SAS},2}$, where the former is responsible for the global

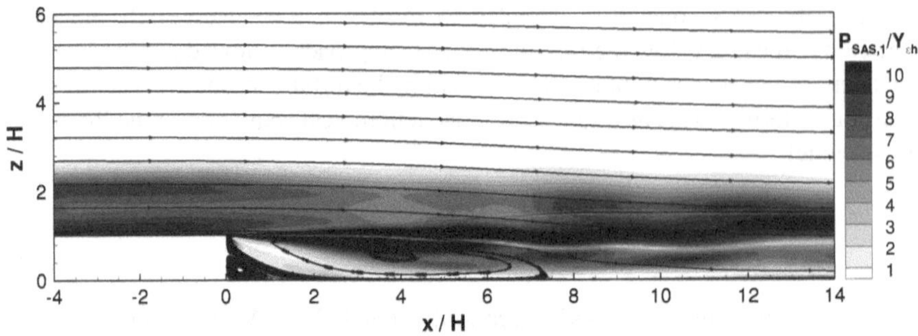

Fig. 1 Backward-facing step. Ratio of $P_{\mathrm{SAS},1}$ to main destruction term Y_{ε^h}

impact of the term compared to the regular source terms. The latter modifies the term itself.

In Fig. 1, the distribution of the ratio of $P_{\mathrm{SAS},1}$ to the main destruction term Y_{ε^h}[1] is shown for the backward-facing step flow, simulated with the basic JHh-v2 RSM. A strong contribution can be seen in the developing shear layer between the recirculating flow and the outer flow. Furthermore significant values are found in the boundary layer upstream of the step, very close to the wall.

Corresponding profiles of $P_{\mathrm{SAS},1}$ at two different streamwise positions, upstream of the step ($x/H = -4$) as well as near the center of the recirculation region ($x/H = 4$), can be seen in Fig. 2.

The second part $P_{\mathrm{SAS},2}$ has its main contribution very close to the wall, as well as in the upper part of the shear layer. As an outcome of Eq. (9) for the sink term, the distribution of $P_{\mathrm{SAS},2}$ is subtracted of the distribution of $P_{\mathrm{SAS},1}$.

As one can see from the skin-friction coefficient in Fig. 3, the basic JHh-v2 model predicts a recirculation zone which is considerably too long. Introducing the first part of the P_{SAS} term with a negative sign into the length-scale equation reduces the turbulence dissipation within the developing shear layer, leading to a shortened recirculation zone. Due to the increased turbulence in the near-wall boundary layer, the skin-friction upstream of the step as well as in the recovering boundary layer downstream of the recirculation zone rises. The intensity of both effects, earlier reattachment and higher skin friction in the boundary layer, can be adjusted by the coefficient $C_{\mathrm{SAS},1}$.

Considering the second contribution to P_{SAS}, the near-wall peak of $P_{\mathrm{SAS},2}$ in Fig. 2 (left) reduces the boundary-layer skin friction, while the reattachment point is only slightly influenced (Fig. 3). With increasing $C_{\mathrm{SAS},2}$, the skin friction is further reduced, simultaneously the influence on the recirculation zone rises. Therefore c_f

[1]Due to the modeling formulation of a combined production-destruction, i.e. derived from two terms in the exact ε equation, the sink term is originally referred to as $P_\varepsilon^4 - Y$ [7]. As a simplification and in order to illustrate its function as destruction term of the ε^h equation, the sink term is in this work denoted as Y_{ε^h}.

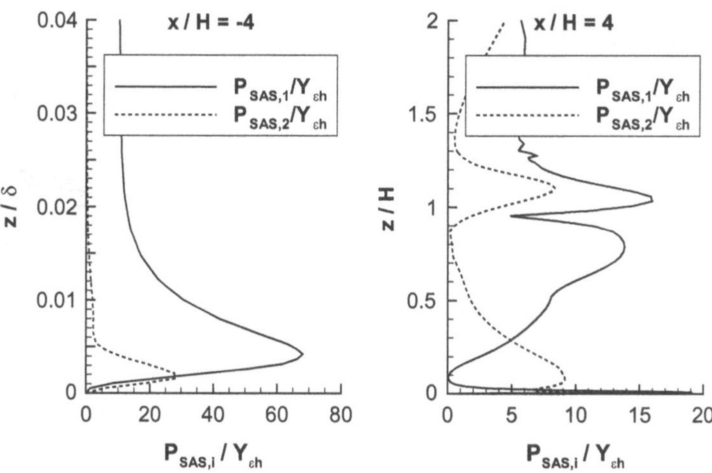

Fig. 2 Backward-facing step. Ratio of $P_{SAS,1}$ and $P_{SAS,2}$ to main destruction term $Y_{\varepsilon h}$; $C_{SAS,1} = 0.008$, $C_{SAS,2} = 2.0$

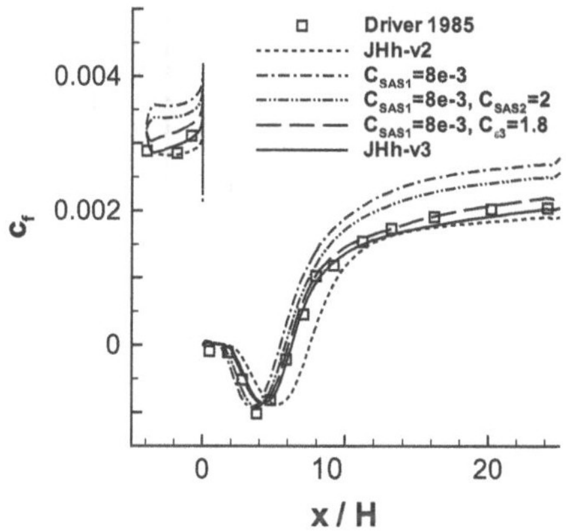

Fig. 3 Backward-facing step. Skin-friction coefficient along the wall

cannot be corrected by modifying $C_{SAS,2}$ only. Within the original model, the low-Reynolds production term $P_{\varepsilon 3}$ with the coefficient $C_{\varepsilon 3} = 0.7$ can be used for this purpose. By increasing the coefficient $C_{\varepsilon 3}$, the skin friction can be reduced.

In Fig. 4, the turbulent shear stress distribution in a profile immediately downstream of the step at $x/H = 1$ can be seen. While the basic RSM drastically underestimates the turbulence in the shear layer, $P_{SAS,1}$ can help to improve the prediction. The amplifying effect of $P_{SAS,1}$ is reduced by increased coefficients for $P_{SAS,2}$ and $P_{\varepsilon 3}$.

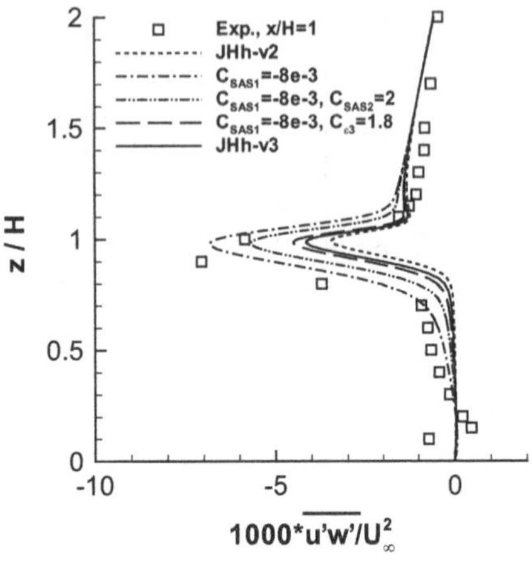

Fig. 4 Backward-facing step. Profile of turbulent shear stress at streamwise position $x/H = 1$

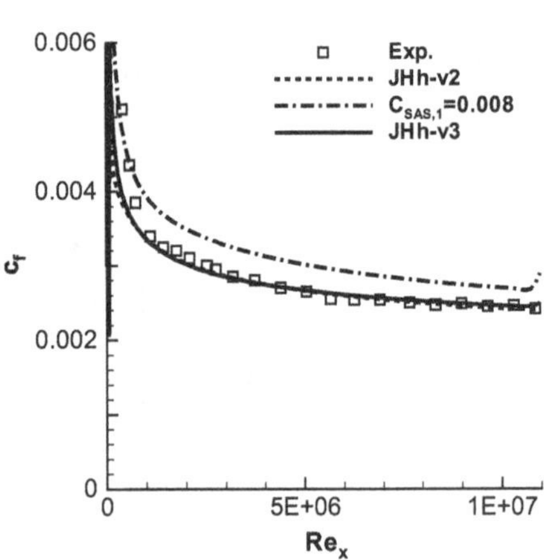

Fig. 5 Zero-pressure-gradient flat plate. Skin-friction coefficient along the wall

For a precise adjustment of the skin friction in boundary layers, a zero-pressure-gradient (ZPG) flat plate is simulated. As we have already seen in the backward-facing step flow, the skin friction rises when $P_{\mathrm{SAS},1}$ is activated (Fig. 5). This effect can again be counteracted by introducing the second part $P_{\mathrm{SAS},2}$ of the sink term, furthermore by increasing $C_{\varepsilon 3}$.

Agreeable results for both calibration cases, backward-facing step as well as ZPG flat plate, are achieved for the following coefficient set: $C_{\mathrm{SAS},1} = 0.008$, $C_{\mathrm{SAS},2} = 2.0$, $C_{\varepsilon 3} = 1.8$. The resulting model is named as JHh-v3. When the shear-stress profile in Fig. 4 is considered, even higher values of $C_{\mathrm{SAS},1}$ seem appropriate to obtain agreeable turbulence levels. The coefficients found here are however regarded as a reasonable compromise of the cases investigated so far.

4 Validation of the Extended Model

Since the model formulation and its coefficients are changed, the new turbulence model version has to be validated against different test cases, in order to evaluate its performance for a wide range of aeronautical flows. In this work, the free shear flow of an axisymmetric single-stream jet is investigated, furthermore the transonic 2D flow around the airfoil RAE 2822 and a transonic axisymmetric bump flow are simulated.

4.1 Round Single-Stream Jet

Similar to the backward-facing step flow, a shear layer with inflection point develops between a jet and the stagnant or slowly moving ambiance. The turbulent $M = 0.75$ jet that emerges from a round nozzle was simulated with both RSM versions, using a nozzle geometry which was experimentally investigated within the EU project JEAN [10] and a 360° hexahedral mesh with 9 million points. The topology of the grid as well as the dimensions of the discretized flow field can be observed in Fig. 6. The grid extends for $15D$ in radial direction and $56D$ in streamwise direction, 180 points are used for circumferential resolution. In order to preserve a high radial grid homogeneity, the polar grid topology is combined with a cartesian topology near the centerline.

Using the nozzle's exit diameter $D = 0.05\,\mathrm{m}$ as reference length, the Reynolds number of the jet is approximately one million. For both RSM simulations the URANS solver was applied, with a physical time-step size of $4 \cdot 10^{-5}$ s and 100 inner iterations. In the initial phase of the simulations, the developing shear layer shows unsteady fluctuations. After a high enough amount of turbulence is produced, the shear layer stabilizes and results in a steady-state solution.

Velocity profiles as well as profiles of the turbulent shear stress component $\overline{u'v'}$ can be seen in Fig. 7, in comparison with experimental results [10]. Already early

Fig. 6 Round single-stream jet. Cut-out of full mesh and cross-section at nozzle exit

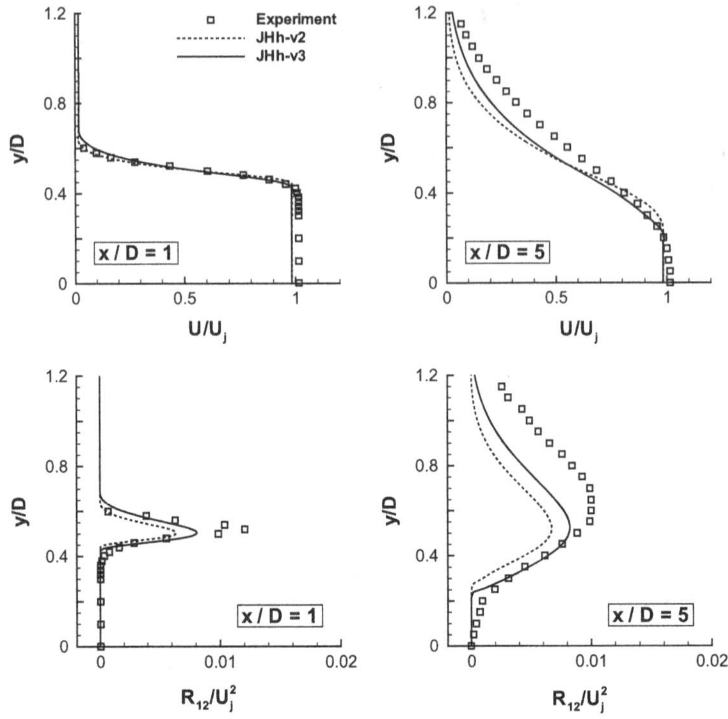

Fig. 7 Round single-stream jet. Profiles of velocity (*upper figures*) and turbulent shear stress (*lower figures*), at two streamwise positions

in the developing shear layer ($x/D = 1$, with x having its origin at the nozzle exit), the JHh-v2 model underestimates the turbulent shear stress, similarly to the BFS flow. The lack of turbulence is still existent at $x/D = 5$, showing furthermore an influence on the velocity profile. An improvement is found in the simulation with the JHh-v3 model, where higher levels of turbulence increase the transport of momentum, giving a velocity profile with a better agreement to the experimental data. Nevertheless the peak values of $\overline{u'v'}$ are underpredicted.

The underestimated transport of momentum has a lengthening effect on the jet core, which can be seen in Fig. 8. The velocity on the jet axis is shown for both RSM versions as well as for the experiments, with U_j being the axial velocity at $x/D = 0$. While in experiments the axial velocity on the centerline starts to decrease at $x_c/D = 5\ldots6$, JHh-v2 predicts $x_c/D \approx 10.4$. An improvement is achieved with the new model version, JHh-v3 simulates a jet core length of $x_c/D \approx 8.8$.

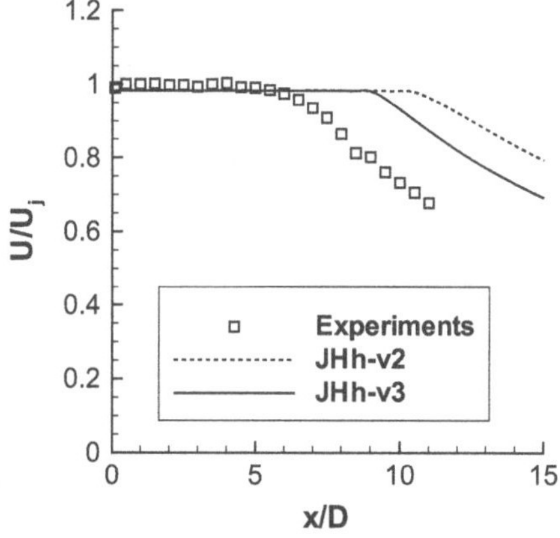

Fig. 8 Round single-stream jet. Axial velocity ratio along the jet axis

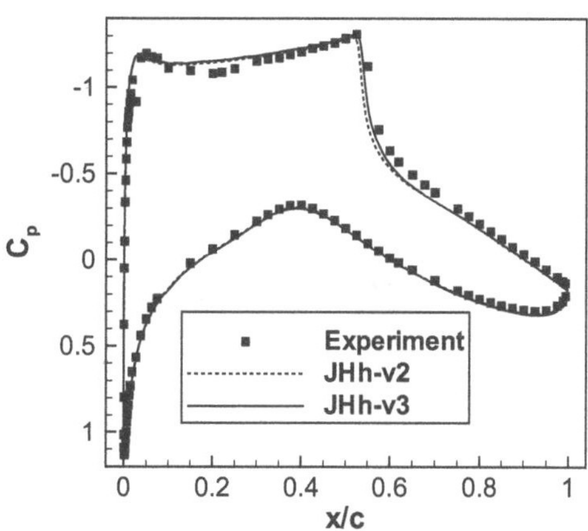

Fig. 9 Transonic airfoil RAE 2822, Case 9. Pressure coefficient along the airfoil

4.2 *Transonic Airfoil RAE 2822*

The supercritical flow around the RAE 2822 airfoil is a standard test case for turbulence models, experimental data is provided by Cook et al. [4]. Figure 9 shows a comparison of measured and simulated pressure distribution at Mach number M $= 0.73$ and Reynolds number Re $= 6.5 \cdot 10^6$, also known as Case 9. The experimental angle of attack is given as $\alpha = 3.19°$, with a recommendation to use $\alpha = 2.8°$ in numerical 2D simulations to consider wind-tunnel-wall interference [4]. It can be seen that the influence of the additional sink term in combination with the re-calibration is rather small for this fully attached airfoil flow. Considering the more critical Case 10 with Mach number M $= 0.75$ and Reynolds number Re $= 6.2 \cdot 10^6$, which develops a shock-induced separation bubble on the upper surface as reported in [4], discrepancies between both Reynolds-stress models can be found around the shock region (Fig. 10). For both cases Fig. 11 displays the ratio of the additional sink term P_{SAS} to the main destruction term Y_{ε^h}, computed

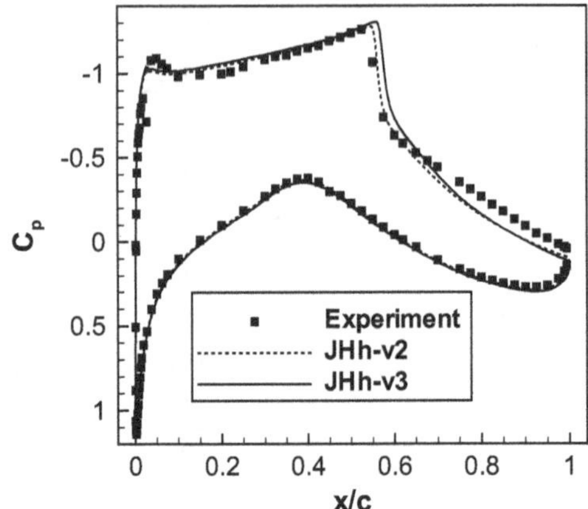

Fig. 10 Transonic airfoil RAE 2822, Case 10. Pressure coefficient along the airfoil

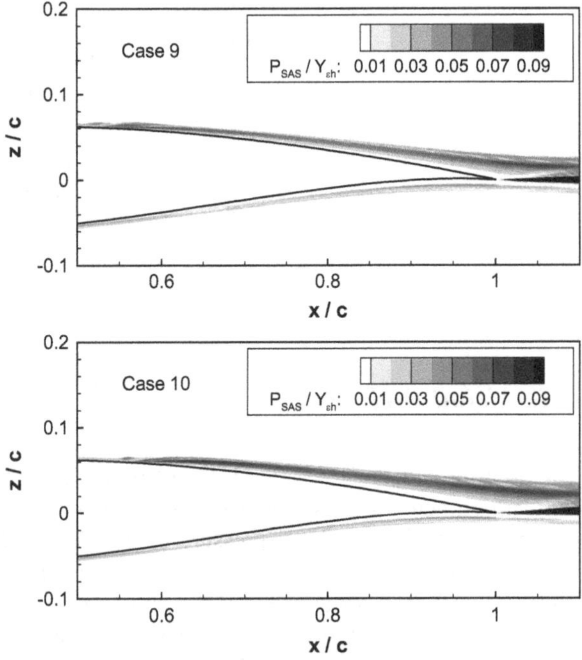

Fig. 11 Transonic airfoil RAE 2822. Ratio of additional sink term P_{SAS} to main destruction term Y_{ε^h}

a priori based on the JHh-v2 results. Except for the wake region, P_{SAS} shows a rather moderate contribution to the ε^h budget, as expected for this mostly attached wall-bounded flow. Slightly higher values are found in the shock region, increasing the turbulence and therefore stabilizing the boundary layer. At the separation in Case 10, an S-shaped velocity profile exists which enhances P_{SAS} compared to Case 9, resulting in a stronger stabilizing effect of the boundary layer. Due to the additional turbulence, the momentum loss in the boundary layer towards the trailing edge decreases, which shifts the shock location downstream. While in Case 9 both RSM versions show a good agreement to the experimental data, the shock location simulated with the JHh-v3 model is positioned slightly downstream in Case 10 compared to the experiments.

4.3 Transonic Axisymmetric Bump

The flow at transonic Mach number over an axisymmetric bump is a complex test case for turbulence models, which was experimentally investigated by Bachalo and Johnson [1, 2]. It develops a compression shock which interacts with the boundary layer and induces a flow separation over the rear part of the bump, before it reattaches further downstream. In order to adequately predict the shock location as well as the separation point, a good quality of the turbulence model in simulating the upstream boundary layer through favorable and adverse pressure gradient is essential. The reattachment point however strongly depends on the turbulent shear stresses that develop in the separated shear layer.

Employing an axisymmetry boundary condition on both sides, the flow was simulated on a 2D grid of 48,000 points that was rotated by 5°. Three hundred points are spent in streamwise direction, of which 150 discretize the bump geometry. The boundary layer upstream of the bump is resolved by 90 points in wall-normal direction.

For evaluating the quality of the simulations, the residual of the transport equation for the Reynolds-stress component R_{11} and the maximum turbulent kinetic energy within the flow field are depicted over the iteration number in Fig. 12 for both RSM versions. Note that well-converged solutions were applied as initial flow field, therefore the residuals are reduced by a rather low number of orders.

Figure 13 shows a comparison of the simulated pressure distribution with experiments. It can be noticed that the basic JHh-v2 model overestimates the size of the separation, which results in an exaggerated pressure plateau around $x/c = 1$. Furthermore the shock position is found slightly upstream of the experimental prediction. Clear improvements of the pressure distribution can be found when using the JHh-v3 model, especially in the recovering boundary layer downstream of the separation ($x/c \approx 1.2$). The pressure plateau is even underestimated, while the shock position shows a good agreement to the experimental data. A comparison of shock position, separation location, and reattachment location is given in Table 1.

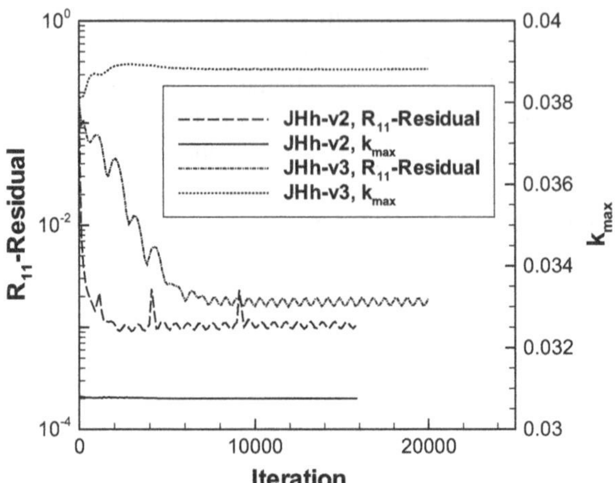

Fig. 12 Transonic axisymmetric bump. Convergence rates of the R_{11}-residual and of k_{max}

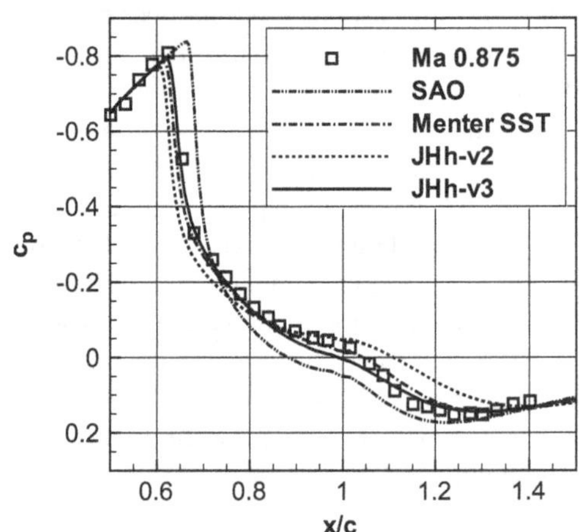

Fig. 13 Transonic axisymmetric bump. Pressure coefficient along the wall

Table 1 Transonic axisymmetric bump. Comparison of simulated and measured flow topology

	SAO	Menter SST	JHh-v2	JHh-v3	Exp.
Shock position[a]	0.690	0.645	0.633	0.653	0.66
Separation point	0.690	0.645	0.667	0.689	0.70
Reattachment point	1.165	1.174	1.194	1.092	1.10
Length of separation	0.475	0.529	0.527	0.403	0.40
Distance of separation to shock	0.000	0.000	0.034	0.036	0.04

[a]In the simulations, the shock position is not a discrete point. The measured shock position corresponds to a pressure coefficient of $c_p = -0.49$, which was used for determination of the simulated shock positions

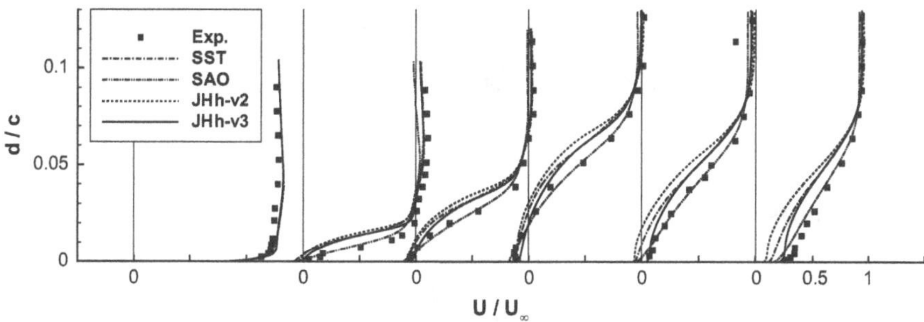

Fig. 14 Transonic axisymmetric bump. Wall-normal velocity profiles in different streamwise positions

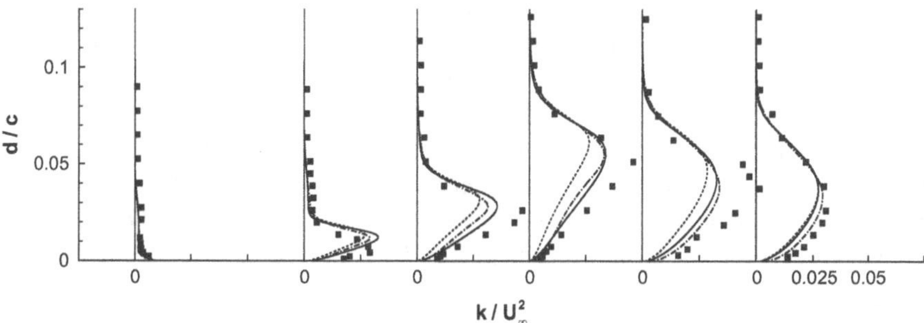

Fig. 15 Transonic axisymmetric bump. Wall-normal profiles of the turbulent kinetic energy in different streamwise positions; legend in Fig. 14

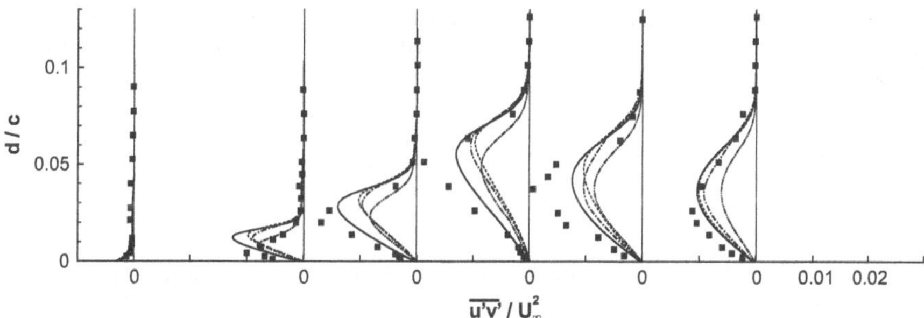

Fig. 16 Transonic axisymmetric bump. Wall-normal profiles of the turbulent shear stress in different streamwise positions; legend in Fig. 14

The prediction of the reattachment point and thus the length of the separation is improved in the simulation with JHh-v3. Furthermore the shock moves slightly downstream, similar to the RAE 2822 case. The reason can be seen in wall-normal profiles of velocity (Fig. 14), turbulent kinetic energy (Fig. 15), and turbulent shear stress (Fig. 16). The momentum loss is reduced in the simulations of JHh-v3, due to higher levels of turbulence in the detaching shear layer.

While both eddy-viscosity models show an immediate separation at the shock, the boundary layer endures the adverse pressure gradient for a short distance in the simulations with both RSM versions, which agrees to the experimental data.

5 Conclusion

The extension of the length-scale equation of the differential Reynolds-stress turbulence model JHh-v2 with an additional sink term in combination with a calibration and validation of the extended model is presented, resulting in the model version JHh-v3. It was shown that the modification amplifies the development of turbulence in free shear flows with large second derivates of the velocity, which positively influences the separation length of a backward-facing step flow. Furthermore the overprediction of the core length in turbulent round jets is reduced. Investigating the flow around the transonic airfoil RAE 2822, only minor influence is noticed in the simulation of the fully attached flow in Case 9. In Case 10 however, the shock position moves slightly downstream, increasing the deviation to the experimental data. In the prediction of the separated flow over a transonic bump, the separation length is reduced with the new model version JHh-v3, leading to a good agreement to experimental data.

Acknowledgements The authors gratefully acknowledge the "Bundesministerium für Bildung und Forschung" who funded parts of this research within the frame of the joint project *AeroStruct* (funding number 20 A 11 02 E), as well as the "North-German Supercomputing Alliance" for supplying us with computational resources within the project *nii00090*. Furthermore we would like to thank A. Probst of DLR Göttingen who provided a computational mesh and his experience for the BFS flow.

References

1. Bachalo WD, Johnson DA (1979) An investigation of transonic turbulent boundary layer separation generated on an axisymmetric flow model. In: AIAA Paper 79-1479, AIAA 12th Fluid and Plasma Dynamics Conference, Williamsburg
2. Bachalo WD, Johnson DA (1986) Transonic, turbulent boundary-layer separation generated on an axisymmetric flow model. AIAA J 24(3):437–443
3. Cécora R-D, Radespiel R, Eisfeld B, Probst A (2014) Differential Reynolds-stress modeling for aeronautics. AIAA J, preprint. doi:10.2514/1.J053250
4. Cook PH, McDonald MA, Firmin MCP (1979) Aerofoil RAE 2822 – pressure distributions, and boundary layer and wake measurements. In: Barche J (ed) Experimental data base for computer program assessment, AGARD-AR-138 A6
5. Daly BJ, Harlow FH (1970) Transport equations of turbulence. Phys Fluids 13:2634–2649
6. Gibson MM, Launder BE (1978) Ground effects on pressure fluctuations in the atmospheric boundary layer. J Fluid Mech 86:491–511
7. Jakirlić S (2004) A DNS-based scrutiny of RANS approaches and their potential for predicting turbulent flows. Habilitation, TU Darmstadt

8. Jakirlić S, Hanjalić K (2002) A new approach to modelling near-wall turbulence energy and stress dissipation. J Fluid Mech 459:139–166
9. Jakirlić S, Maduta R (2014) On "Steady" RANS Modeling for improved Prediction of Wall-bounded Separation. In: AIAA Paper 2014-0586, 52nd AIAA Aerospace Sciences Meeting, National Harbor
10. Jordan P, Gervais Y, Valière J-C, Foulon H (2002) Final results from single point measurements. Project deliverable D3.4, JEAN - EU 5th Framework Programme, G4RD-CT2000-00313, Laboratoire d'Etude Aérodynamiques, Poitiers, France
11. Maduta R (2014) An eddy-resolving Reynolds stress model for unsteady flow computations: development and application. Dissertation, TU Darmstadt
12. Maduta R, Jakirlić S (2012) An eddy-resolving Reynolds stress transport model for unsteady flow computations. In: Fu S et al. (eds) Advances in hybrid RANS-LES modelling 4. Notes on numerical fluid mechanics and multidisciplinary design, vol 117. Springer, Heidelberg, pp 77–89
13. Menter FR (1994) Two-equation Eddy-viscosity turbulence models for engineering applications. AIAA J 32(8):1598–1605
14. Menter FR, Egorov Y (2005) A scale-adaptive simulation model using two-equation models. In: AIAA Paper 2005-1095, 46th AIAA Aerospace Sciences Meeting and Exhibit, Reno
15. Menter FR, Egorov Y (2010) The scale-adaptive simulation method for unsteady turbulent flow predictions. Part 1: theory and model description. Flow Turbul Combust 85(1):113–138
16. Probst A (2013) Reynoldsspannungsmodellierung für das Überziehen in der Flugzeugaerodynamik. Dissertation, Technische Universität Braunschweig
17. Probst A, Radespiel R (2008) Implementation and extension of a near-wall reynolds-stress model for application to aerodynamic flows on unstructured meshes. In: AIAA Paper 2008-770, 46th AIAA Aerospace Sciences Meeting and Exhibit, Reno
18. Reuß S, Probst A, Knopp T (2012) Numerical investigation of the DLR F15 two-element airfoil using a Reynolds stress model. In: Third Symposium "Simulation of Wing and Nacelle Stall", Braunschweig
19. Rotta JC (1972) Turbulente strömungen. B. G. Teubner, Stuttgart
20. Schwamborn D, Gardner A, von Geyr H, Krumbein A, Lüdeke H, Stürmer A (2008) Development of the TAU- Code for aerospace applications. In: 50th NAL International Conference on Aerospace Science and Technology, Bangalore
21. Spalart PR, Allmaras SR (1994) A one-equation turbulence model for aerodynamic flows. La Recherche Aèrospatiale 1:5–21

Zeitfracht Medien GmbH
Ferdinand-Jühlke-Straße 7
99095 Erfurt, Deutschland
produktsicherheit@kolibri360.de